**Green Energy and Technology**

More information about this series at http://www.springer.com/series/8059

Nilesh Y. Jadhav

# Green and Smart Buildings

## Advanced Technology Options

Springer

Nilesh Y. Jadhav
Energy Research Institute
Nanyang Technological University
Singapore
Singapore

ISSN 1865-3529 ISSN 1865-3537 (electronic)
Green Energy and Technology
ISBN 978-981-10-1000-2 ISBN 978-981-10-1002-6 (eBook)
DOI 10.1007/978-981-10-1002-6

Library of Congress Control Number: 2016952823



Printed on acid-free paper

This Springer imprint is published by Springer Nature
The registered company is Springer Nature Singapore Pte Ltd.
The registered company address is: 152 Beach Road, #22-06/08 Gateway East, Singapore 189721, Singapore

*To my loving wife, Dipti, as she sacrificed the most while I tried to balance several commitments while finding time to write this book.*

# Preface

A safe, comfortable, and sustainable built environment is highly desirable as we spent most of our time in offices, factories, or homes. With the rapid urbanization movement throughout the world, it is predicted by the United Nations that 66 % of the population will live in urban areas or cities by 2050. Urbanization can be seen as a boon as well as a key threat to world sustainability. While on the one hand it offers opportunities such as better health, greater access to social services, and enhanced living options, it may also lead to rapid sprawl, pollution, and environmental degradation, if not planned well. It is hence essential to focus on urban sustainability and a sustainable built environment, with green and smart buildings as one of the most important keys to sustainable urban living.

While sustainability and sustainable development is a very broad topic, it is often viewed as consisting of three main elements: economic development, social development, and environmental protection. These three elements are equally important and often intricately linked to each other. It is noted that 'green' is sometimes mistakenly associated with the overzealous and inordinate support and advocacy of the aspect of environmental protection, while completely ignoring economic and social development aspects of sustainable development. This is not, however, the purpose of using the word 'green' in the title of this book. Although environmental protection will appear to be the primary goal of the technology solutions discussed in this book, the economic and social aspects of development and adoption of these technologies cannot be ignored.

The word 'smart' is also an overused term these days with the advent of smartphones and other smart devices. These devices enable users to take automatic control of their surrounding conditions in lightning speed without much human intervention. Smart buildings of the future would incorporate business objectives, user preferences, and sustainability goals seamlessly in building operations, while providing an exceptional experience to the building occupants and other stakeholders.

The specific focus of this book is on buildings. Understandably, buildings are part of a larger ecosystem of building estates and cities. However, the focus of this book is on the building itself and how technologies can help enhance its overall sustainability performance. Building systems cover very broad topics, and there are in-depth publications and books on each of these topics separately. The attempt here is to provide an overview of advanced technology options for green and smart buildings without going into in-depth discussions in each area.

As an author of this book, I sincerely hope that this book will provide valuable reference and motivation on adopting, developing, and further improving through research, the effectiveness of green and smart building technologies. Along with many of the challenges facing humanity today, climate change is an important one and the building sector could do it's part by adopting green and smart buildings. With the wide range of technologies and their benefits described in this book, green and smart buildings should become a norm without any further excuses. I definitely wish that my children and the next generation in general should live, work, learn, and play in buildings that are good for them, good for the economy, and good for the environment.

Singapore Nilesh Y. Jadhav

# Contents

# Chapter 1
# Introduction

The building sector is the largest contributor to global greenhouse gas (GHG) emissions. Buildings use about 40 % of global energy, 25 % of global water, 40 % of global resources, and they emit approximately 30 % of GHG emissions. These emissions are set to double by 2050 if we carry on business as usual (UNEP 2014). The energy used by the building sector continues to increase, primarily because new buildings are constructed faster than old ones are retired. Electricity consumption in the commercial building sector doubled between 1980 and 2000, and is expected to increase another 50 % by 2025 (EIA 2005).

Other important facts:

- The building sector is estimated to be worth 10 % of global GDP (USD7.5 trillion) and employs 111 million people.
- Residential and commercial buildings consume approximately 60 % of the world's electricity.
- Energy consumption in buildings can be reduced by 30–80 % using proven and commercially available technologies.

Figure 1.1 shows that the buildings present the most impactful and also economical mitigation potential for GHG emissions globally.

Greenhouse gas emissions from buildings primarily arise from their consumption of fossil-fuel based energy, both through the direct use of fossil fuels and through the use of electricity which has been generated from fossil fuels. Significant greenhouse gas emissions are also generated through construction materials, in particular insulation materials, and refrigeration and cooling systems. Broadly speaking, energy is consumed during the following activities:

1. manufacturing of building materials ('embedded' or 'embodied' energy)
2. transport of these materials from production plants to building sites ('grey' energy);
3. construction of the building ('induced' energy);
4. operation of the building ('operational' energy); and
5. demolition of the building (and recycling of their parts, where this occurs).



N.Y. Jadhav, *Green and Smart Buildings*, Green Energy and Technology,
DOI 10.1007/978-981-10-1002-6_1

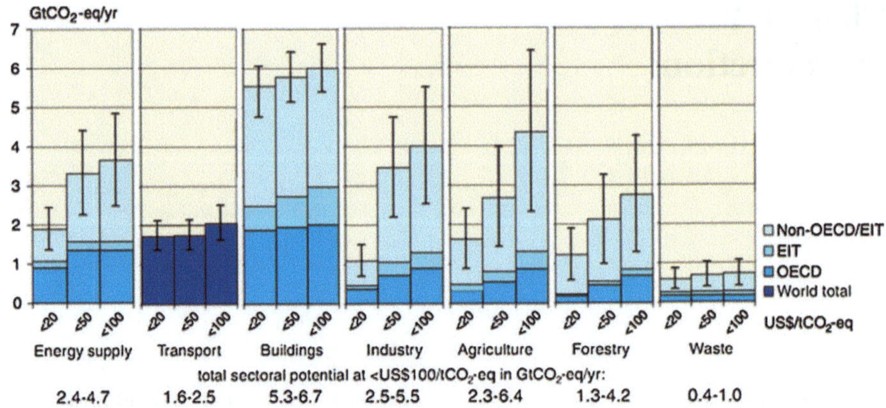


**Fig. 1.1** Estimated economic mitigation potential by sector using technologies and practices expected to be available by 2030 (IPCC 2007)

Though figures vary from building to building, studies suggest that over 80 % of greenhouse gas emissions take place during the operational phase to meet various energy needs such as Heating, Ventilation, and Air Conditioning (HVAC), water heating, lighting, entertainment and telecommunications (Junnila 2004; Suzuki and Oka 1998; Adalberth et al. 2001). A smaller percentage, generally 10–20 %, of energy is consumed in materials manufacturing and transport, construction, maintenance and demolition.

Given these facts, it's imperative that significant actions be taken towards adoption and growth of green buildings, which can lead to significant reduction in GHG emissions especially during the operational phase of the building. Green buildings are designed to reduce the overall impact of the built environment on human health and the natural environment by:

- Efficiently using energy, water, and other resources
- Protecting occupant health and improving productivity
- Reducing waste, pollution and environmental degradation.

There are various definitions of a 'Green Building' and how it performs compared to any other building. One of the most comprehensive definition that I have come across and that will be used as a guideline for this book is as follows:

> Green building is the practice of creating structures and using processes that are environmentally responsible and resource-efficient throughout a building's life-cycle from siting to design, construction, operation, maintenance, renovation and deconstruction. This practice expands and complements the classical building design concerns of economy, utility, durability, and comfort. (US EPA 2014)

The design of building is an important factor affecting its eventual operational performance. Once designed and built, the potential for energy savings is locked and subsequent retrofits or refurbishments can be highly uneconomical. In this book, we look at several green building design technologies, including the design methodology itself (i.e. integrated design). Broadly speaking, green building design features can be divided into two main categories: passive and active design features.

The 'passive design' features impact the energy performance of the building in an indirect way, but at the same time can lead to major reduction in energy consumption of buildings in the operational phase. Passive design takes advantage of natural energy flows to maintain occupant comfort (thermal, ventilation and lighting) instead of overly relying on mechanical and electrical systems for the same. However, the passive design features are often not customisable to suit the dynamic operational conditions such as changes in weather and varying occupancy levels or heat sources. The elements of passive design are:

- Building orientation e.g. with respect to the sun-path
- Form factor of the building: massing, shape
- The building envelope viz. roof, facade, walls, windows, shading devices, etc.
- Insulation and exterior/interior finishes.

The passive design methodologies and technologies will be covered in details in Chap. 4.

The 'active design' features impact energy consumption in building directly and are often easy to customise (tune-up or tune-down) according to the operational requirements. They comprise of the mechanical and electrical provisions such as:

- Air-conditioning (heat/cooling)
- Mechanical ventilation
- Artificial lighting
- Elevators and escalators
- Pumps and compressors.

The active design technologies depend to a large extent on the equipment design and operating efficiency. As they need to be dynamically controlled, there are often moving parts or switching features that need to be maintained regularly in order to maintain high operating efficiencies and thermal comfort. We will review active design technologies for green buildings in Chap. 5.

As active and passive design features can both be impactful ways to reduce the overall energy consumption in buildings, there needs to be an optimal mix of the two considering the overall costs, utility, thermal comfort and reliability. This calls for an integrated design approach where various requirements, constraints and conditions are considered with the goal to design a green building. We will look at integrated design in details in Chap. 3.

Considering the fact that operational energy is the highest life-cycle contributor in buildings, it is important that building operations be cognizant to this fact. Building owners, facility managers and occupants should be aware and empowered to take actions necessary to reduce operational energy without sacrificing comfort and utility of the building. This brings us to the topic of 'Smart Buildings'.

A smart building can be defined as follows:

> Smart buildings use information technology during operation to connect a variety of subsystems, which typically operate independently, so that these systems can share information to optimize total building performance. Smart buildings look beyond the building equipment within their four walls. They are connected and responsive to the smart power grid, and they interact with building operators and occupants to empower them with new levels of visibility and actionable information. (Institute of Building Energy Efficiency 2011)

A basic but important element of a smart buildings is the Building Management System (BMS) or Building Automation System (BAS). There are many commercially available BMS and BAS systems for buildings and they consist of the following essential elements:

- Data acquisition: sensors, meters, etc.
- Database management
- Analytics
- Visualisation
- Controls.

Each of the above are relevant technology development areas with the goal to make the systems more powerful and cost-effective in operational energy savings. Such systems also ensure that the building is performing to its design standards at all times and trigger operation and maintenance activities to deal with deviations in performance. The Building Management and Automation technologies are covered in details in Chap. 6.

Apart from providing living and working spaces with good occupant comfort and operational functionality, buildings can be also looked upon as avenues for distributed energy generation especially using renewable energy technologies. These technologies include Solar Photovoltaic (PV)/Solar Thermal energy generation, Wind Energy and Geothermal Energy. Each of these technologies are, on their own, going through a lot of technological developments and often a challenge is to integrate them seamlessly in the building without compromising safety, comfort and functionality. We cover this topic in more details in Chap. 7.

Other than consuming energy in various forms (e.g. thermal and electric), buildings also consume water for several activities such as the provision of basic amenities like toilet and wash-rooms, cooking, landscaping, cleaning and also a significant amount of water can be consumed in the air-conditioning systems such as cooling towers. Buildings also produce waste such as paper, plastic, glass, food, cans, electronic waste and office equipment and furniture during its operational phase. Apart from this, there is construction waste during its construction and demolition phases. From recycling and reuse to purposeful at-source reduction, there are several strategies to reduce the water and waste in buildings. These strategies are covered in Chap. 8.

The business case for green and smart buildings is often fragmented and not strong due to the fact that the stakeholders viz. owner, developer, builder, operator and occupant, each have a role to play. Often, a well-designed green and smart building is not able to achieve its desired environmental performance due to misunderstanding or lack of understanding between the occupants or the users of the green features provided in the design. There are evolving techniques such as green-lease, green occupancy handbooks and incentive schemes that are useful to emphasise proper operations and sustainable utilisation of green and smart building features. We look at such occupant engagement techniques and tools in Chap. 9.

Green and Smart Building rating systems play an important role in benchmarking and assuring better performance. After all, how do we know if enough is being done to be able to classify a building as a green or smart building. There are now several green building rating systems throughout the world and each of them differ in how certain design and operational aspects are assigned weightage in the overall rating calculation. In Chap. 10 we look at various green and smart building performance assessment and rating systems and how they can impactfully shape the green building industry if used in an appropriate way.

Lastly, we look at two green and smart building case study examples in the concluding Chap. 11. There are now numerous such examples around the world and these examples prove that green and smart buildings should be aspired to as a norm and not as an exception or a high reaching ideal.

Thus a green and smart building is an interplay of various integrated design strategies such as passive and active design features, as well as water and waste reduction techniques, renewable energy integration, building management systems and controls, efficient operations and rating systems that allow effective benchmarking and performance analysis and guidelines for various stakeholders involved in the building industry. Figure 1.2 summarizes the essential elements of a green and smart building.

A green and smart building uses both technology and process to create a facility that is environmentally friendly, cost-effective and intelligence driven, while being safe, healthy and comfortable for its occupants. It provides its facility managers and owners with tools that allow for timely and intelligent decision making during its design, operations and maintenance phase. It enhances the user experience through logic and intelligence that caters for changing occupant requirements, climatic conditions and maintenance needs, while optimizing its performance throughout the life-cycle of the building.

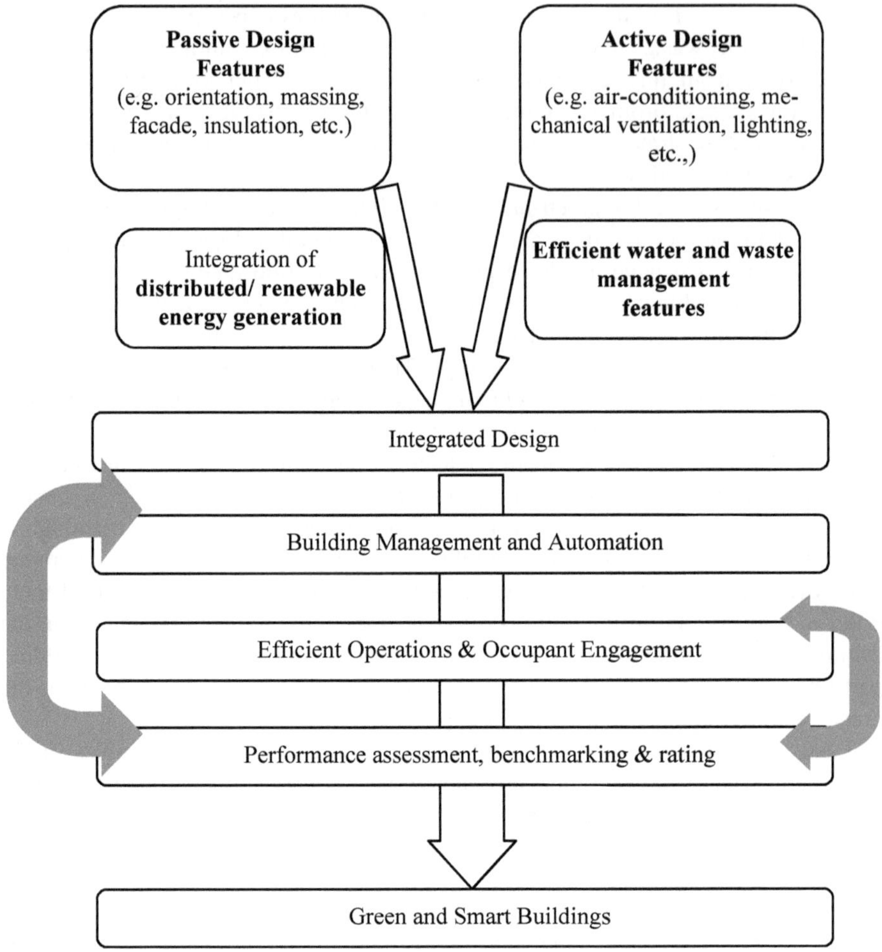


**Fig. 1.2** Essential elements of a green and smart building

## References

Adalberth K, Almgren A, Petersen EH (2001) Life cycle assessment of four multi-family buildings. Intl J Low Energy Sustain Build 2:1–21

EIA (2005) Annual energy review 2004. www.eia.doe.gov/emeu/aer/contents.html. U.S. Department of Energy, Energy Information Administration, Washington, DC

Institute of Building Energy Efficiency (2011) http://www.institutebe.com/smart-grid-smart-building/What-is-a-Smart-Building.aspx

IPCC (2007) Climate change 2007: synthesis report. In: Core Writing Team, Pachauri RK, Reisinger A (eds) Contribution of working groups I, II and III to the fourth assessment report of the intergovernmental panel on climate change. IPCC, Geneva, Switzerland, 104 pp

IPCC (2013) Climate change 2013: the physical science basis. In: Working group I contribution to the fifth assessment report of the intergovernmental panel on climate change

Junnila S (2004) An environmental impact of an office building throughout its life cycle. Doctoral Dissertation, Helsinki University of Technology Construction Economics and Management, Espoo

Suzuki M, Oka T (1998) Estimation of life cycle energy consumption and $CO_2$ emission of office buildings in Japan. J Energy Build 28:33–41

UNEP (2014) http://www.unep.org/sbci/AboutSBCI/Background.asp

US EPA (2014) http://www.epa.gov/greenbuilding/pubs/about.htm

# Chapter 2
# Green and Smart Building Trends

Green and Smart Buildings have evolved over the past two decades from being an environmental idealism to a business case with strong industry growth and a long-term business opportunity. Most of the building construction sector stakeholders such as architects, engineers, planners, owners and consultants are now increasingly focusing on green building projects in their portfolio. This is driven by factors such as client/market demand, lower operating costs, greater environmental consciousness, regulatory requirements/incentives and improved health and productivity benefits. Construction of green buildings rose to 325 million m$^2$ of new floor space in 2013, representing a $260 billion market, according to a report by Lux Research (Lux Research 2014).

## 2.1 Business Case for Going Green

According to the latest report by Dodge Data and Analytics (formerly McGraw Hill Construction), the global green building activity continues to double every three years. The percentage of firms expecting to have more than 60 % of their projects certified green is anticipated to more than double from 18 % currently to 37 % by 2018. It is encouraging to note that the growth of green building activity is also happening in emerging economies such as countries in Asia and Africa. The top triggers for architects and engineers to focus on green buildings are client demand, environmental regulations, market demand and the notion of 'right thing to do'. However, there are several barriers or obstacles for green buildings such as higher perceived first costs, lack of public awareness, lack of political support and incentives, and the perception that green is for the high-end project sector only (Dodge Data and Analytics 2016).



N.Y. Jadhav, *Green and Smart Buildings*, Green Energy and Technology,
DOI 10.1007/978-981-10-1002-6_2

The report also highlights several recognized business benefits of green buildings such as follows (in order of decreasing importance based on sector survey):

1. Lower operating costs (savings in energy and total life-cycle costs)
2. Quality assurance (provided by the documentation and certification)
3. Education of occupants about sustainability
4. Higher building value at point of sale
5. Future proofing of assets
6. Flexibility of design built into green buildings
7. Increased productivity for tenants
8. Higher rental rates
9. Higher occupancy rates.

A study conducted by Lux Research (Lux Research 2014) covered utility savings, rental rates, resale value and government incentives in green buildings and their impact on internal rate of returns. The following are their main findings:

- Energy efficiency codes offer unparalleled market opportunity for green buildings. While green building standards and certifications helped build market demand for green buildings, building energy efficiency codes can create a much larger market opportunity. In Germany, Lux estimates that in 2013, new floor space compliant with the EnEv 2009 code was 50 million $m^2$, or about 36 % of overall new construction.
- Green certification fuels growth and buildings with higher level of certification standards outperform their baseline peers. As an example, the report notes that higher rental income added $4.1 million in value to a model 80,000 $ft^2$ commercial building in Los Angeles.
- Internal rates of return are increased from subsidies. Incentives like Germany's subsidized interest rates for energy-efficient homes, or government cash rebates in India, can lead to an internal rate of return of 5–6 % over 15 years.

## 2.2 Smart Buildings Solutions on the Rise

Smart building solutions are attractive tools for energy management because they provide visibility into system-wide operational and energy efficiency and provide tools for analyzing, tracking, and communicating the impact of energy-efficiency improvements. According to a new IDC Energy Insights report (IDC 2015), smart building technology spending will grow from $6.3 billion in 2014 to $17.4 billion in 2019, registering a compound annual growth rate of 22.6 %. The most aggressive adoption will be in Asia/Pacific, North America, and Western Europe.

There is an increasing availability of building data than can be processed through big data analytics tools to understand possibilities of making building greener and smarter. There is also applications of analytics engines and cognitive computing to analyse and interpret the data to detect anomalies and identify opportunities to optimize building performance at all times. For the US, estimates predict that smarter buildings could save US$ 20–25 billion in annual energy costs. This opportunity is largely untapped today, as many building owners and operators are not yet aware of how data-driven optimizations can reduce energy consumption (Accenture 2011).

The adoption of smart building solutions in North America, Europe and Asia/Pacific is particularly positive due to a combination of policy and business dynamics, which help drive market awareness of the business value of these solutions and as a result the market demand. There are also growing regulations at the local, regional, and national levels driving awareness of the benefits of energy efficiency and energy benchmarking. In addition, a number of large corporations have made commitments to sustainability and energy management.

## 2.3 Green and Smart Buildings Standards and Regulations

Governments across most major economies of the world are starting to care about energy efficiency and sustainability in the building sector. This has resulted in several regulations, bylaws, codes and standards related to green buildings. Some governments have introduced clear incentives and sanctions related to green building performance. Although the level of activity and intervention from government bodies varies from region to region, it is clear that there is a clear positive shift towards and smart green buildings globally.

There are also various non-government bodies that are actively championing the growth and proliferation of green building standards and implementation. One such well known and highly active non-government entity is the Green Building Council. Green building councils are member-based, non-profit organisations that empower industry leaders to effect the transformation of the local building industry toward sustainability. With one hundred thousand buildings and almost 1 billion $m^2$ of green building space registered, the influence and impact of this global network is a significant force for social and environmental change. The World Green Building Council is a network of national green building councils in more than one hundred countries, making it the world's largest international organisation influencing the green building marketplace (please refer to their website: http://www.worldgbc.org for further details).

## 2.4 Financing Green and Smart Buildings

Another useful trend for the green building industry in the financing realm is the model of a Energy Services Company (ESCO) that provide innovative financing schemes along with the full suite of energy conservation measures. The initial or capital costs of implementing the energy conservation measures are financed through the energy savings achieved by the measures over a pre-determined period of operation of the building. Another slight variation of this is a 'performance based contract' where the ESCO only gets paid fully if the pre-determined energy savings are achieved through the implementation of energy savings measures. These mechanisms are useful for retrofitting or refurbishing existing buildings into green and smart buildings.

## 2.5 Zero Energy Buildings

The next frontier in Green and Smart Buildings is the move towards Zero energy buildings. Energy efficiency solutions have already become mainstream in many regions around the world. Some building developers are now pushing the limits and trying to differentiate themselves by going towards net zero energy buildings. A net zero-energy building (ZEB) is a residential or commercial building with greatly reduced energy needs through efficiency gains such that the balance of energy needs can be supplied with energy generated locally from renewable technologies. In this respect, the use of solar photovoltaic (Solar PV) technology on buildings for local renewable energy generation is gaining prominence. With the cost of solar PV components dropping exponentially over the last decade and availability of third party financing schemes, the adoption rate of solar PV in buildings has been increasing rapidly. Other technologies for on-site renewable energy generation include solar hot water collectors, small-scale wind power turbines and geothermal energy.

## 2.6 Buildings Enabled by IoT

The advent of Internet of Things (IoT) is also affecting the way green and smart buildings are getting connected with various stakeholders throughout the world. IoT for buildings can be imagined as a network for buildings that are embedded with electronics, software, sensors, and network connectivity that enables them to collect and exchange data. In 2016 and beyond, building owners and key decision makers will invest in an array of smart building solutions that embody the technology foundation of Internet of Things (IoT) and cloud computing, according to a new white paper from Navigant Research.

Rapid deployments of cost-effective data acquisition devices and the integration of IT and traditional building automation and controls are changing the industry of facilities management. This network-addressable IT backbone is generating continuous data streams associated with building operations. An evolving menu of software analytics and energy management services is leveraging this data to provide actionable insights for efficiency and optimization. The result is the development of intelligent buildings that are dynamic systems capable of managing energy consumption for efficiency and a whole suite of non-energy business improvements.

The whitepaper lists the following 10 trends that reflect the evolving facilities industry and the push for more intelligent buildings (Navigant Research 2016):

1. Building energy management systems (BEMS) will deliver benefits beyond energy efficiency. Building Energy Management System Revenue Is Expected to Reach $10.8 billion in 2024.
2. The energy cloud will redefine buildings as energy assets.
3. Climate change policy will mandate efficiency in buildings.
4. Intelligent buildings will optimize the occupant experience.
5. Innovative financing opportunities will disrupt the market.
6. Cyber security will be a key differentiator between intelligent building solutions.
7. Hardware and software analytics will improve health in buildings.
8. Intelligent building water management will emerge as a priority for innovators.
9. Smart cities will drive the development of intelligent buildings.
10. Electric utilities and energy providers will ramp investment in BEMSs.

## 2.7 People Centric Focus in Buildings

Buildings are also increasingly becoming people centric rather than engineering, construction or technology centric. The employee expectations are rising constantly for better technology in the workplace. This is leading to building owners, developers and facility managers to focus on providing more collaborative spaces, flexible and adaptable spaces, and increase their focus on occupant well-being and comfort. There is an increasing focus on building smarter workplaces that engage occupants, understand how they use buildings in new ways and get them involved in implementing sustainable practices.

## 2.8 Green Buildings at the Paris Climate Conference 2015

The world experienced a historical agreement on climate change mitigation action in the Paris Climate Conference in December 2015. It was for the first time in over 20 years of United Nations (UN) negotiations, that the Conference of Parties

(COP) was able to achieve a legally binding and universal agreement on climate, with the aim of keeping global warming below 2 °C.

The buildings sector's crucial role in mitigating climate change was also being recognised. For the first time in the history of climate negotiations, a Buildings Day was held on 3rd December 2015 at the Paris Climate conference. On this occasion, the Global Alliance for Buildings and Construction was launched to effectively to combat Climate Change. The alliance is supported by the United Nations Environment Programme and has been joined by 20 States, representing more than 1 billion people. It also has participation of 8 major corporations and over 50 national and international organisations.

The Global Alliance for Buildings and Construction (Global ABC) will work to facilitate mobilisation of international resources for efficient local operational solutions, align existing initiatives, commitments and programmes to achieve greater scale, and catalyse a greater pace and impact of climate action in the buildings and construction sector. The following are the objectives highlighted for this global alliance (UNEP 2016):

- Bring together all the relevant global players on a large scale around a common ambition and sustain this momentum to ensure that they work together over time;
- Increase the share of green building in international funding to implement new initiatives and increase the visibility of exemplary initiatives;
- Gather around a program of operational activities strategic networks and partners covering the full range of stakeholders in the building production chain;
- Promote initiatives and solutions by all the members signatory to the Alliance to make them reproducible and ensure their appropriation;
- Create a network for public authorities in charge of construction, to align regulations and financing towards low-carbon strategies.

## References


Accenture (2011) Energy-smart buildings—demonstrating how information technology can cut energy use and costs of real estate portfolios. Whitepaper. Available online from: http://czgbc.org/energy-smart-buildings-whitepaper.pdf. Retrieved 12 Dec 2014

Dodge Data and Analytics (2016) SmartMarket report—world green building trends. Available online from: http://fidic.org/sites/default/files/World%20Green%20Building%20Trends%202016%20SmartMarket%20Report%20FINAL.pdf. Retrieved 28 Feb 2016

IDC Energy Insights (2015) Business strategy: global smart building technology spending 2015–2019 forecast

Lux Research (2014) Cash is king: assessing the financial performance of green buildings. State of Market Report, Lux Research Inc.

Navigant Research (2016) Intelligent buildings: 10 trends to watch in 2016 and beyond. Navigant Research Publication

UNEP (United Nations Environment Programme) (2016). Available online at: http://web.unep.org/climatechange/buildingsday/about-us. Accessed 14 Mar 2016

# Chapter 3
# Integrated Design Concepts and Tools

It is critical to make the decision to build a green building early in the design and construction, or retrofitting and refurbishment process in order to maximize the green potential, minimize redesign, and assure the overall success and economic viability of the green elements of the building project. Making a commitment to build green and establishing firm environmental objectives for the project must be done as early as possible because opportunities for incorporating green technologies and design solutions become less and less available and increasingly costly to implement as the project design and construction process progresses.

Building a green building is not just a matter of assembling a collection of the latest green technologies or materials. Neither is it sufficient to just follow the building codes, standards or rating system guidelines that are available for green buildings. It is essential to ensure that choices made for green building technologies and measures are commensurate with the site selection, building orientation in order to effectively meet the design goals. There are various factors to be considered during the design of a green and smart building as shown in the Fig. 3.1.

## 3.1 Integrated Design Process

Integrated design or whole-building design is a process in which every element of the design is first optimized and then the impact and interrelationship of various different elements and systems within the building and site are re-evaluated, integrated, and optimized as part of a whole building solution. For example, interrelationships between the building site, site features, the path of the sun, and the location and orientation of the building and elements such as windows and external shading devices have a significant impact on the quality and effectiveness of natural day lighting. These elements also affect direct solar loads and overall energy performance for the life of the building. Without considering these issues early in the design process, the design is not fully optimized and the result is likely to be a very



N.Y. Jadhav, *Green and Smart Buildings*, Green Energy and Technology,
DOI 10.1007/978-981-10-1002-6_3

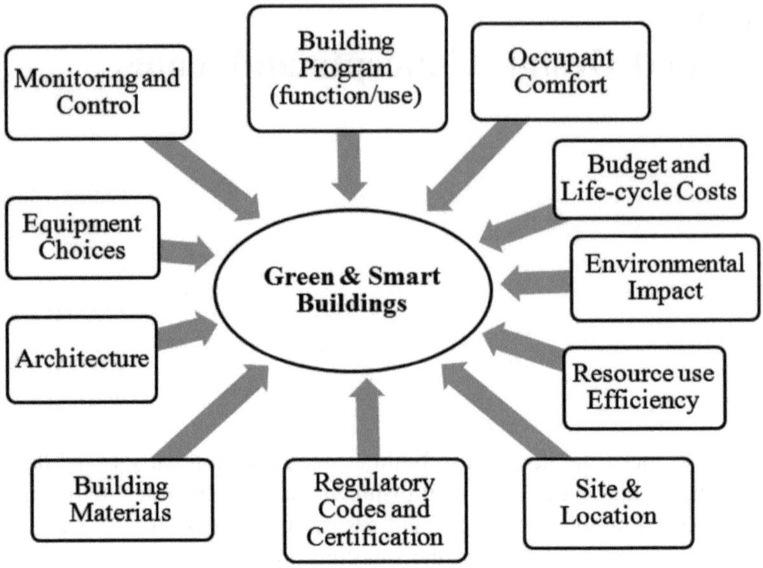


**Fig. 3.1** Factors affecting green and smart building design

inefficient building. This same emphasis on integrated and optimized design is inherent in nearly every aspect of the building from site planning and use of on-site storm water management strategies to envelope design and detailing and provisions for natural ventilation of the building.

The integrated design process mandates that all of the design professionals work cooperatively towards common goals from day one. This is often easier said than done as each professional and stakeholder have their own perspective, goals and performance targets for the building design and quite of a few of them could be competing or contradicting with each other. However, by establishing communication and collaboration early in the project, there will be less "surprises" that cause delays and have higher cost implications. Table 3.1 highlights the differences between a traditional building design approach and an integrated design approach.

One useful method to enhance the communication between the various stakeholders of the building is to conduct a joint workshop or what is often referred to as a 'Integrated Design Charrette'. A charrette can be viewed as a focused and collaborative brainstorming session held at the beginning of a project with a multi-disciplinary team that encourages an exchange of ideas and information and allows truly integrated design solutions to take form. Participants are encouraged to cross fertilize and address problems beyond their field of expertise. The charrette is particularly helpful in complex situations where many people represent the interests of the client and conflicting needs and constituencies. Participants are educated about the issues and resolution enables them to 'buy into' the schematic solutions. A final solution isn't necessarily produced, but importantly, often interdependent,

**Table 3.1** Difference between a traditional and integrated design approach

| Traditional design approach | Integrated design approach |
|---|---|
| The project team members are involved sequentially based on their professional specialization | A diverse set of inputs from all project stakeholders is sought right from the start of the project |
| Project intensity is increased over time with less time spent at early design stages | The intensity of project in the early stages is significant with brainstorming discussions, design charrettes and planning workshops/meetings |
| The building systems are considered and optimized in isolation, often resulting in over-sizing due to design buffers | The project team adopts a whole system thinking approach considering overall building performance |
| Energy modeling and building simulation is done towards the end to assess performance and/or meet regulatory requirements | Energy modeling and simulation is done on a continuous and iterative basis throughout the design process to meet performance goals optimally |
| There is often a strong focus on reducing upfront capital costs | A life-cycle cost approach is taken to achieve better total cost savings throughout the building life cycle |

issues are explored. The following are the desired outcomes from a successful Integrated Design Charrette (NREL 2009):

- Establish a multidisciplinary team that can set and agree about common project goals. See Fig. 3.2 on the areas of expertise desired from the multidisciplinary team.
- Develop early consensus about project design priorities.
- Set project performance goals.
- Generate quantifiable metrics to measure the final energy and environmental outcomes against the performance goals.
- Provide early understanding of the potential impact of various design strategies.
- Initiate an integrated design process to reduce project costs and schedules and obtain the best energy and environmental performance.
- Identify project strategies to explore with their associated costs, time considerations, and needed expertise to eliminate costly surprises later in the design and construction processes.
- Identify partners, available grants, and potential collaborations that can provide expertise, funding, credibility, and support.
- Set a project schedule and budget that all team members feel comfortable following.

Figure 3.3 shows a photo from a real design charrette that indicates the participatory nature of such workshop, which is crucial in achieving its success.

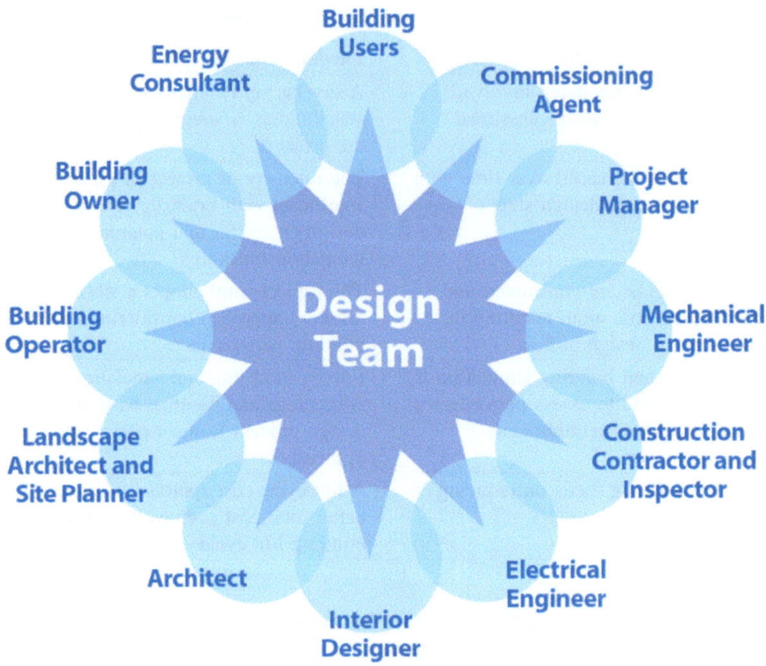


**Fig. 3.2** Areas of expertise to be represented in a Integrated Design Charrette (NREL 2009)

**Fig. 3.3** Photo taken during an Integrated Design Charrette (ERI@N, Singapore 2013)

## 3.2 Building Information Modeling

Tools such as Building Information Modeling (BIM) are very useful during the design stage of the building to get a fairly accurate visual representation of the building and document its design features in an easily accessible manner. BIM is a is a digital 3D model of all aspects of a building including architectural, structural and mechanical elements. BIM has the potential to change the role of drawings for the construction process, improve architectural productivity, and make it easier to consider and evaluate design alternatives. BIM also aids in the process of integrating the various design professionals to find problems like ducting and wiring systems that are drawn in the same location, further encouraging and demanding an integrated design process.

BIM is a substantial enhancement over the previously used Computer Aided Design (CAD) tools in terms of functionality and most importantly allowing collaborative designs. It differs from the previously popular architecture drafting tools by allowing the addition of further information such as time, cost, manufacturers' details, sustainability and maintenance information, etc. to the building model. With everyone working within one file it is easy to see what information is missing and what areas of the design need to be focused on or refined and hence significantly aids the decision making process. See Fig. 3.4 for a screenshot from a BIM tool used on a computer.

There are various commercially available BIM software programs and the following list shows some popular BIM software programs and providers:

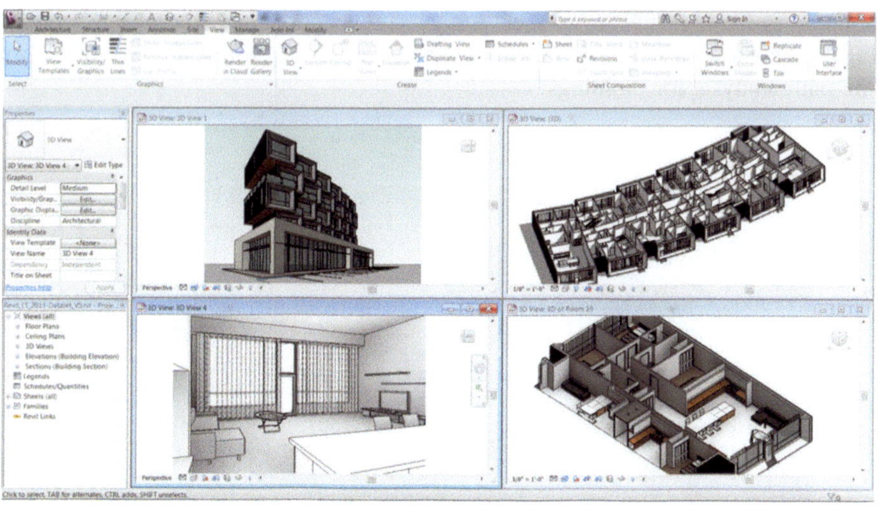

**Fig. 3.4** Screenshot from a BIM software tool

1. Autodesk provides various BIM software programs such as Revit for architectural design, Mechanical, Electrical and Plumbing (MEP), and structural engineering as well as cloud based solutions such as BIM360. Visit for more information: http://www.autodesk.com/.
2. Bentley Systems also provides a suite of BIM tools: https://www.bentley.com/.
3. DProfiler by Beck technology: http://www.beck-technology.com/.
4. ArchiCAD by Graphisoft: http://www.graphisoft.com/.
5. Vectorworks Designer by Nemetschek: http://www.nemetschek.net/.
6. Tekla Structures: http://www.tekla.com/.
7. Google Sketchup Pro: http://www.sketchup.google.com/.
8. Affinity by Trelligence: http://www.trelligence.com/.
9. Vico Office: http://www.vicosoftware.com/.
10. Digital Project by Gehry technologies: http://www.gehrytechnologies.com/.

## 3.3 Energy Modeling and Simulation

Energy modeling and simulation tools allow the design team to evaluate the energy performance of the building in the virtual environment considering its geometrical model, design parameters, occupancy schedules and local weather conditions. The output of the energy modeling is typically the energy consumption of the building simulated at time intervals (typically hourly) throughout the year. It could be used as a tool to compare and contrast various technologies for energy efficiency of the building at the design stage. This will allow the design team to choose appropriate building technologies vis-a-vis their energy savings potential and cost implications. Figure 3.5 shows a screenshot from the energy modeling software used to perform sun path analysis for buildings.

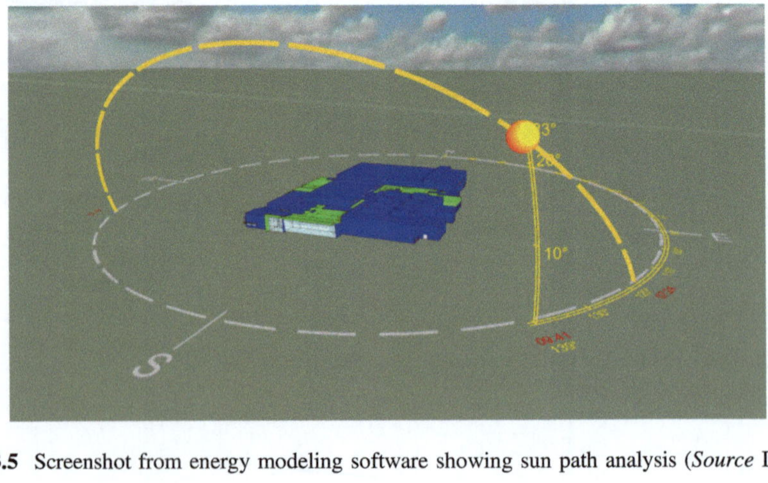


**Fig. 3.5** Screenshot from energy modeling software showing sun path analysis (*Source* IES)

**Table 3.2** Benefits of energy modeling at various stages of building development (adapted from AIA 2012)

| Building stage | Energy modeling outputs |
|---|---|
| Concept design | ✓ Define goals and requirements of the project based on early modeling results<br>✓ Experiment with options for building orientation, massing and envelope constructions<br>✓ Explore ways to reduce loads<br>✓ Get the design team together to assess integrated system performance and make design decisions based on the results |
| Schematic design | ✓ Create a baseline energy model<br>✓ Test the effectiveness of different energy conservation measures and options before selection and implementation<br>✓ Determine the most cost effective solution to reduce energy consumption to the lowest |
| Design development | ✓ Finalize baseline model to compare against proposed design<br>✓ Create proposed models with system alternatives to choose from<br>✓ Provide annual energy use charts and other performance metrics for comparing baseline versus proposed design<br>✓ Evaluate specific products and technologies with refined data and details |
| Construction documents | ✓ Create documentation needed to accompany energy model results for code compliance and for green certification requirements such as LEED |
| Commissioning and post-occupancy | ✓ Use results of the as-built model for commissioning<br>✓ Collect metered operating data to create a calibrated model to share with outcome-based database |

Although energy modeling is primarily used in the design stage, it can also be very useful in the other stages of the building development. Table 3.2 highlights the benefits and outputs of energy modeling at different stages of development of the building. Design, performance, and energy modeling are iterative processes. Initial models address fundamental design parameters, including the building envelope, orientation and massing, typically without including mechanical or electrical systems in a manner that provides crucial, and sometimes surprising, design guidance. As models develop, they provide feedback to the design team on how the form, orientation, programmatic strategies, and other variables will likely affect the project's performance in terms of energy, daylighting, comfort, and other design characteristics. E.g. it is useful to perform a dynamic shading analysis of the building structure to understand the ingress of heat and daylight into the building at various times and seasons during the year. This facilitates the choice of building orientation, form factor and envelope parameters.

It is important to note that Building Energy Modeling is a computer program and the accuracy of results would largely depend on the inputs provided to the model. In this context, one should be mindful of the expression 'garbage-in-garbage-out'. It is hence crucial that the right set of inputs are provided to the energy modeling tool. Table 3.3 lists the typical inputs provided to the energy modeling tool. Some

**Table 3.3** Typical input parameters required for building energy modeling

| Model input set | Input parameters |
|---|---|
| Location specific | • Local climate data (typically imported via a weather file in the software)<br>• Interior conditions and set points |
| Architectural massing and form | (Typically imported through 3D geometrical modeling tools such as Google Sketchup)<br>• Building shape and orientation<br>• Total floor area<br>• Number of floors and thermal zoning<br>• Floor-to-ceiling height |
| Building envelope | • Window-to-wall ratio<br>• Area, orientation, solar absorptance, and thermal transmittance of all opaque building surface<br>• Area, orientation, solar heat gain coefficient, visible light and thermal transmittance, and shading of all glazing components<br>• Mass of building components<br>• Infiltration rates |
| Thermal and electric loads | • Lighting intensity<br>• Plug loads intensity<br>• Sensible and latent (moisture) loads from people and other equipment<br>• Pumps, motors, fans, elevators |
| Schedule of operations | • Lighting schedules<br>• Plug-load schedules<br>• Occupancy schedules |
| Mechanical and Electrical (M&E) systems | • Cooling/Heating system type, including the source, distribution, and terminal units<br>• Ventilation system type<br>• Fan and pump inputs<br>• Economizers and/or heat recovery systems<br>• Domestic hot-water system<br>• Specialty systems (e.g. fume hoods, exhausts)<br>• Renewable-energy systems |

user-friendly software programs have built-in industry standard defaults of these input parameters that speed up early model creation.

The past decade has seen remarkable growth in the Building Energy Modeling industry, primarily driven by more stringent building energy efficiency standards and a growth in voluntary building energy certification programs. There are now various energy modeling and simulation tools that are commercially available to perform the functions as described above and the popular ones are listed below for quick reference:

1. EnergyPlus by the U.S. Department of energy (free tool), available at: https://energyplus.net/.
2. IES Virtual Environment (VE): http://www.iesve.com/.
3. eQUEST, quick energy simulation tool: http://www.doe2.com/equest/.
4. Autodesk Green Building Studio: https://gbs.autodesk.com/GBS/ (cloud based).

5. TRNSYS: http://www.trnsys.com/.
6. DesignBuilder: http://www.designbuilder.co.uk/.
7. Radiance: Lighting simulation tool: http://www.radiance-online.org/.
8. OpenStudio by NREL: https://www.openstudio.net/ (open source, free interface tool using EnergyPlus and Radiance.
9. Dymola: http://www.3ds.com/products-services/catia/products/dymola.
10. Autodesk Revit (previously known as Ecotect).

## 3.4 Advance Optimization Tools

It can be extremely difficult to optimize a building's energy consumption or cost-effectiveness due to the vast array of parameters that can be changed in the energy model during the design phase. To maximize the energy-savings potential while minimizing the costs of a building, there are advance tools available such as Opt-E-Plus (from NREL) and GenOpt (from Berkeley Lab). Such tools automatically run thousands of simulations and identifies design options that provide the most economical energy savings. The design options are based upon parameters chosen by the user that are allowed to vary within specified ranges, and all possible combinations are plotted on a chart of percentage energy savings plotted against percentage cost savings in order to outline optimal design strategies. An example results chart is shown in the Fig. 3.6.

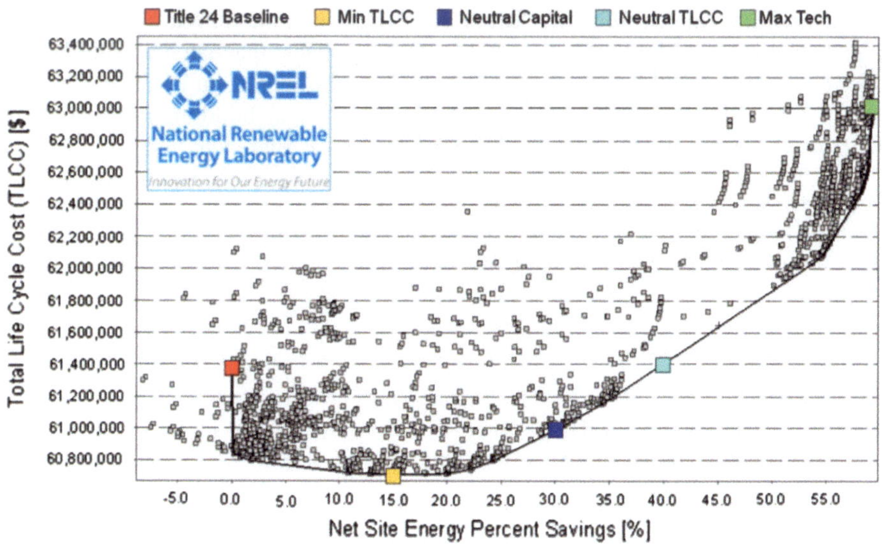


**Fig. 3.6** Example results from NREL's Opt-E-Plus advanced optimization tool for energy modeling (NREL 2009)

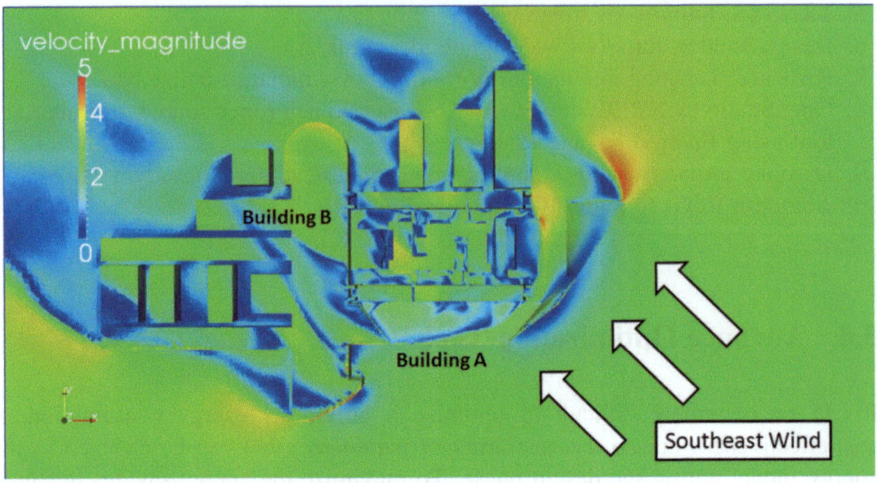


**Fig. 3.7** Screenshot from Computational Fluid Dynamics (CFD) modeling software used for airflow analysis across neighborhood buildings (ERI@N, Singapore)

## 3.5 Airflow Modeling

Apart from energy modeling, it is sometimes important to understand the air flow and distribution patterns for buildings: both outside and inside. The building form and shape can affect how air flows through the building and across neighboring developments into the building. This is an important consideration for natural ventilation and can significantly reduce costs of air-conditioning provisions. There are Computational Fluid Dynamics (CFD) tools available that can help simulate the air-flow patterns within built-spaces as well as for whole building estates (see Fig. 3.7 for illustration of airflow modeling analysis in buildings). The popular tools used for this purpose are listed below for quick reference:

- Fluent by Ansys: http://www.ansys.com/.
- FloVent from Mentor Graphics: http://www.mentor.com/.
- Comsol Multiphysics modeling software: https://www.comsol.com/.

## References

AIA (The American Institute of Architects) (2012) An Architect's guide to integrating energy modeling in the design process

ERI@N (Energy Research Institute @ NTU) (2013) Nanyang Technological University (NTU), Singapore

NREL (2009) A handbook for planning and conducting charrettes for high-performance projects, National Renewable Energy Laboratory (NREL), Sept 2009

# Chapter 4
# Passive Design Technologies

## 4.1 Introduction

The term 'Passive Design' here refers to design strategies, technologies and solutions that effectively take advantage of the environmental conditions outside the building to maximise the energy and cost savings while ensuring the core building facilities and provisions (such as indoor comfort, safety, health, etc.) are not compromised. The environmental conditions can provide several advantages or disadvantages to the building such as the following:

- *Day lighting*: can reduce the energy used for artificial lighting but excessive and improper exposure may result in glare and other forms of visual discomfort
- *Natural ventilation*: can reduce mechanical ventilation energy to move air around but can result in hygiene issues and over-cooling in cold climates
- *Natural cooling*: to reduce the need for excessive air-conditioning or mechanical cooling in hot climates
- *Natural heating*: to use the energy from the sun to provide heat indoors in cold climates instead of providing excessive artificial heating. But this needs to be managed in hot climates to reduce air-conditioning energy use
- *Shading*: (from trees or neighbouring buildings) can reduce heat from direct sun exposure in hot climates but can obstruct views and natural light and heat in cold climates.

The key elements of passive design are:

1. Building location and orientation on the site
2. Building layout and massing
3. Building envelope (windows, walls, roof, insulation and shading).

These passive design elements (as summarized in Fig. 4.1) are massive levers to achieve a high performance building in the early stages of the building design process. Once fixed, it's often very difficult and costly to change these elements as



N.Y. Jadhav, *Green and Smart Buildings*, Green Energy and Technology,
DOI 10.1007/978-981-10-1002-6_4

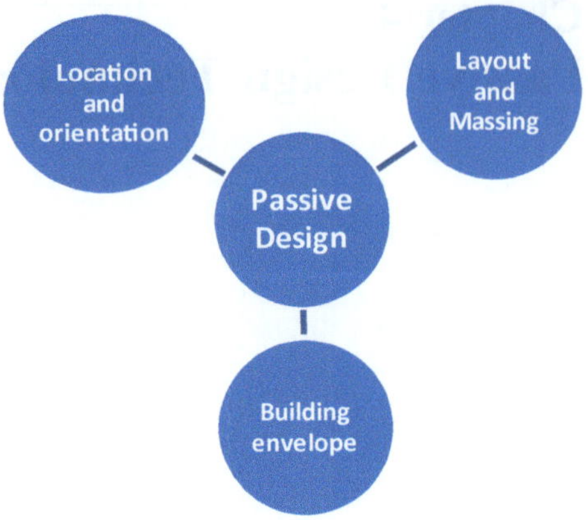


**Fig. 4.1** Key elements of passive design for green buildings

the building development progresses. After construction it is almost impossible to change some of these elements such as the orientation and massing without having a drastic and dramatic impact on the project costs and timeline. Let us consider these elements in more details and understand the associated key terminology and parameters before jumping into the technologies.

## 4.2 Building Location and Orientation

The location of the building on the site and its orientation can have a significant impact on its performance, considering that it can impact the following:

- The amount of sun (light and heat) a building receives
- The intensity of wind (breeze and ventilation)
- Access to views
- Access to transportation options and other site provisions
- Impact on or from neighbouring buildings.

The amount of sunlight and heat (solar heat gain) received by the building is often a key factor in determining the orientation of a building (see Fig. 4.2). Depending on the geographical location and outside weather conditions, this might mean different things for different designs. e.g. in cold climates with mild summers, it might be desirable to get more sunlight and solar heat gain in order to reduce the use of artificial lighting and heating. Whereas, in a tropical weather, its bets to avoid solar heat gain in order to reduce the amount of energy used for air-conditioning. However, it might be also desirable to harness the sun for solar energy generation.

It is also important to consider the surrounding site topography and elevation as that affects the wind conditions. e.g. a building at higher elevation is likely to have higher and more consistent wind conditions and buildings on the leeward side of a hill could produce stagnant wind conditions irrespective of the orientation

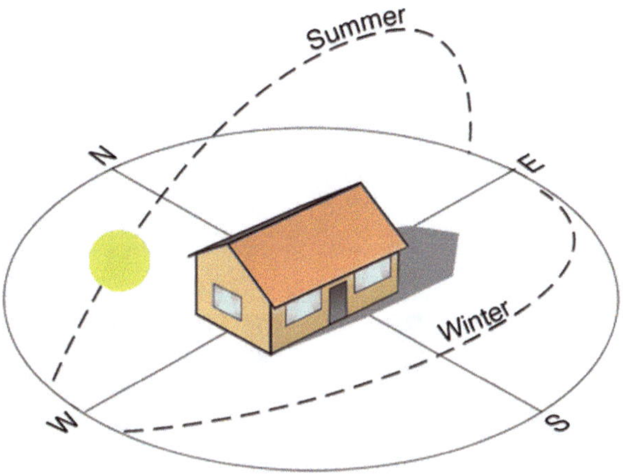


**Fig. 4.2** Considering the sun path to decide building orientation (Autodesk 2015)

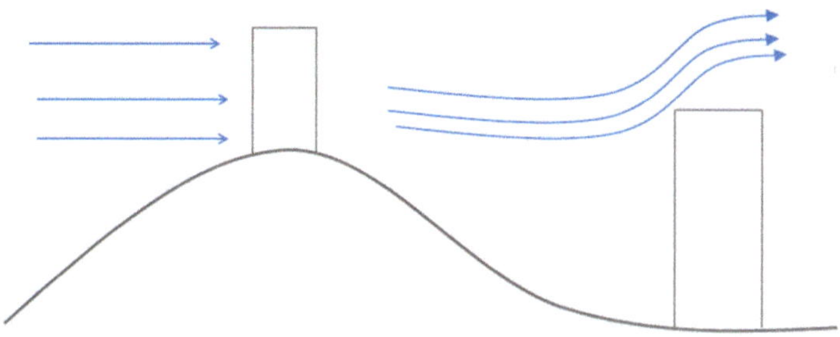

**Fig. 4.3** Location of building resulting in different wind conditions

(see Fig. 4.3). The wind conditions can have a significant impact on energy use of the building especially in hot weather conditions where a naturally ventilated building can achieve comfortable conditions indoors without the excessive need for cooling or air-conditioning. The wind considerations (wind speed and direction) would also affect the building massing and location of certain equipment within the building considering the wind draft and wind load.

## 4.3 Building Layout and Massing

"Massing" is the architectural term used when determining the overall layout (e.g. compact or spread-out), shape (e.g. square, rectangle, oval), size (e.g. height, length, width) and form factor (more solid or porous with cut-outs) of the building. Massing choices depend on the project specifics such as the project site and its goals. It should be optimized in the early stages of the building along with its

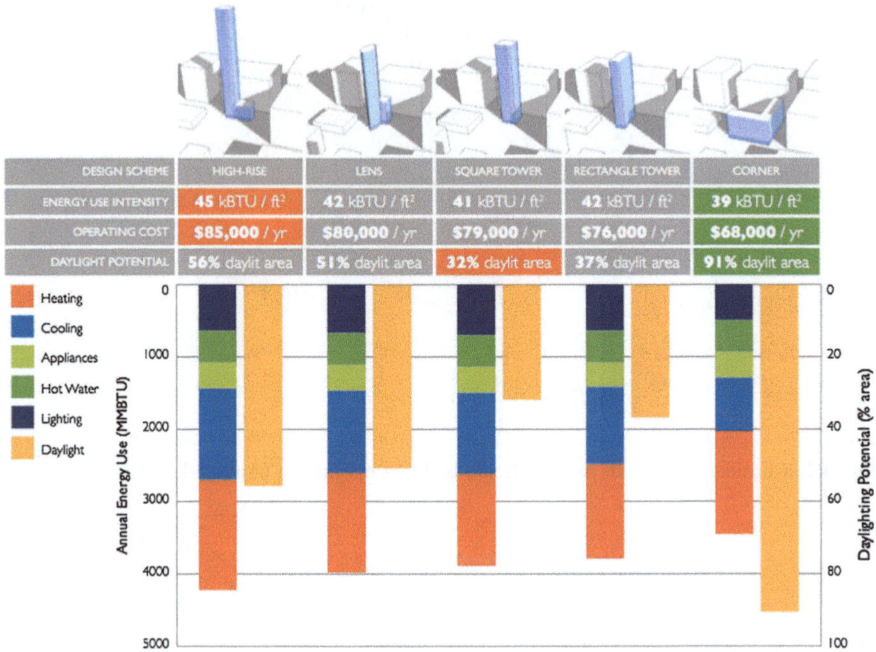


**Fig. 4.4** Massing options and implication on building performance (Sefaira 2013)

location and orientation. Successful massing for green buildings would help to reduce energy loads for the building by leveraging on the natural effects such as the wind flow patterns and the sun path.

Apart from reducing energy use, a good massing and orientation can also help in rainwater harvesting, integrating the building with urban planning provisions, reducing material usage and in protecting ecologically sensitive elements such as greenery, trees and ponds. Modelling and simulation tools as described in the earlier chapter can be used to assess the performance of the building with various orientation and massing choices. Figure 4.4 shows the impact of different massing options on the building's energy use intensity, operating costs and daylight potential.

## 4.4 Building Envelope

The building envelope is the 'skin' of the building. In simpler terms it is the is the physical separator between the exterior and the interior built environment (see Fig. 4.5 for a simple illustration of the building envelope). The key components of the building envelope are as follows:

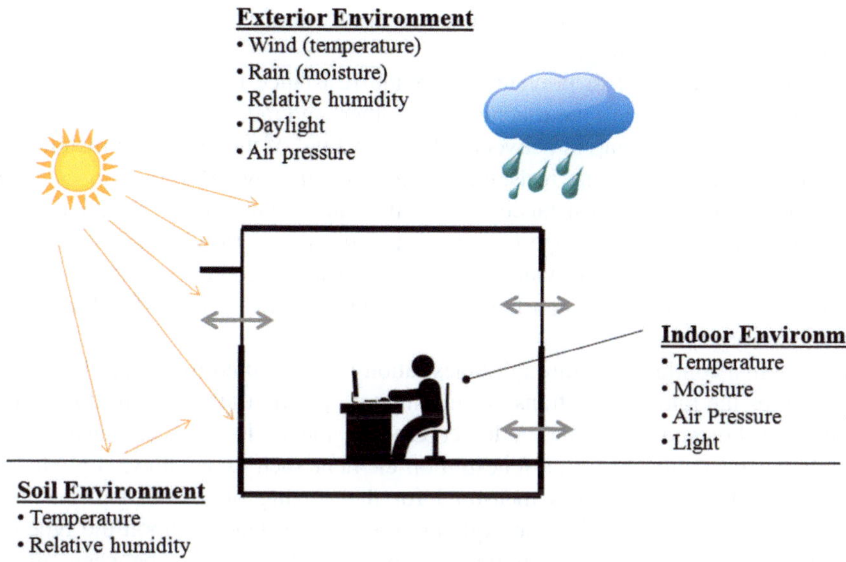


**Fig. 4.5** Building Envelope and its interaction with the outdoor environment

- Roof: the covering on the uppermost part of a building
- Wall: the vertical structure connecting the roof and floor
- Floor: the bottom surface of the building or the walking surface
- Fenestration: openings in the structure such as windows, skylights and doors. When that opening is covered with a translucent or transparent surface (like glass), it's called glazing.

The choice of the building envelope is governed by the climate, culture, aesthetics and available materials. It has a significant impact on the performance of the building as it can determine the amount of sun light/heat and air ingress in the building. The choice of materials used for the building envelope also largely determines its carbon footprint and environmental friendliness.

In order to assess the performance of the building envelope, it is important to understand the material properties of the building envelope components such as their insulation, thermal resistance, heat gain coefficient, air infiltration and visual light transmittance.

### *4.4.1 Insulation*

Insulation is the material that provides resistance to the transfer of heat through the building envelope. Insulation performs a critical function in building design because it enables spaces to avoid excess heat gain from outside and to retain the heat they have inside. Thus is helps to passively influence the heating or cooling energy used by the building.

Insulation is designed to prevent or provide resistance to heat transfer due to conduction (when materials are touching each other) and radiation (when there is no direct contact but an air-gap in between). The 'R-value' of the insulation material is a parameter that measures its resistance to conduction (high R-value means high thermal resistance). The resistance of the insulation material to radiative heat transfer is measured by its property termed as 'emissivity' (low emissivity means high resistance and high reflectance of radiative heat). A good insulation material used for a green building design hence should ideally have a high R-value and a low emissivity.

The thermal conductance through fenestration is a measured using the 'U-value' that specifies the rate of heat transfer per unit area per unit temperature difference between the hot and cold side. A lower U-value means lower heat transfer and hence better insulation from the fenestration element such as window, skylight or glazing. U-value is sometimes measured for the glazing only or it can be also specified for the entire window or skylight assembly. It depends upon the type of glass used for the glazing, the number of panes and air-gap in between, glass tinting, and reflective coatings.

### *4.4.2 Solar Heat Gain Coefficient (SHGC)*

The Solar Heat Gain Coefficient (SHGC) is a ratio of the heat transmitted through the building envelope into the building to the heat that is reflected away. SHGC is a dimensionless parameter and can theoretically range from zero to one, with one representing that all of the heat being transmitted through the envelope, and zero representing none of the incident solar heat being transmitted inside. A low SHGC would be desirable for buildings with high cooling load (e.g. tropical and hot climates), whereas a high SHGC would be beneficial for buildings with passive heating requirements (e.g. cold climate conditions).

### *4.4.3 Infiltration or Air Leakage*

Infiltration or air leakage is caused by air entering or leaving the building due to unintentional gaps in the building envelope. The outside air infiltrated into the inner space would cool or heat interior spaces through convection and make the envelope insulation redundant. The air leakage from inside to outside would result in energy wastage and overload on air treatment facilities such as heating or cooling as the treated air would escape to the surroundings instead of heating or cooling the space.

Infiltration or air leakage can be a problem for the overall building performance as it directly affects heating and cooling requirements in buildings. The severity of its impact however depends on the surrounding climatic conditions.

In cold climates, where the temperature difference between outside and inside can be significant, infiltration of outside air can cause huge energy penalties for the building. In hot climates, as the temperature difference between the outside and inside is low, infiltration will only moderately affect the energy performance. In moderate climates, infiltration can in fact be beneficial due to the natural ventilation caused by the infiltrated air that can reduce the energy for cooling, especially for spaces with high internal loads.

### *4.4.4 Visible Light Transmittance (VLT)*

The percentage of visible light that passes through a window or glazing material is characterized by the parameter known as the Visible Light Transmittance (VLT). An opaque wall would have a zero VLT (0 %), whereas an unobstructed and empty facade opening would have a 100 % VLT. This property only measured the light in the visible portion of the spectrum (and not infrared light). A properly designed glazing unit with high VLT can reduce the electric lighting load and its associated cooling load.

## 4.5 Building Envelope Technologies

### *4.5.1 Shading*

Shading are simple envelope attachments that can prevent the heat and glare caused by direct sunlight through windows. They are also helpful in reducing the direct sunlight incident on walls and roofs so that the building design is not overly reliant on insulation or high performance glazing in hot climates. Shading fixtures can be either provided inside the space or applied from outside and each arrangements have their own benefits and disadvantages.

#### 4.5.1.1 External Shading Fixtures

An exterior fixture such as the horizontal overhangs (see Fig. 4.6) or vertical fins are the most common shading devices used in buildings. While horizontal overhangs are good for cutting of overhead sun exposure, vertical fins are useful in avoiding low-angled sun. Sun shading systems can be designed so as to provide great architectural impact as well as being highly functional. Shading systems now

**Fig. 4.6** Photo of a building with exterior horizontal shading fixtures (Levolux 2012)

come in a great variety of materials such as glass, metal, wood, acrylic and fabric louvers and the choice depends on the effectiveness in the given climate and their architectural integration.

The shading dimensions (height and width) can be calculated in such a way as to allow the sunlight and heat into the building at particular time of the day in the year (e.g. during winter when the sun angle is low), while rejecting it at other times (e.g. during summer when the sun angle is high). Both these scenarios would minimise the heating and cooling loads of the building. This is the benefit of smart passive building design through the apt use of weather and climatic information coupled with simple envelope fixtures.

#### 4.5.1.2 Motorized Internal Shading Devices

Internal shading devices such as roller or Roman shades or venetian blinds are commonly found in homes and commercial spaces and occupants can control or operate them easily. It can also be motorized and movable with automatic controls that are based on weather conditions and sunlight sensors or on fixed daily operating programs. Motorised and automatic internal shading can be an effective way to adapt to external conditions while ensuring occupant comfort.

### *4.5.2 Cool Roofs and Coatings*

Cool roofs use materials with high solar reflectance (high albedo) that are able to reflect the sunlight that a conventional roof would otherwise absorb as heat. The heat absorbed by the roof would result in indoor discomfort or higher costs for cooling in warm climates. While insulation slows the transfer of heat into the building due its high resistance to conduction (high R-value), it does not eliminate the source of the heat gain. Cool roofs can reduce the need for excessive insulation in hot climates by enhancing the ability of the roof to reject solar heat. Figure 4.7 shows the effect of cool coating applied on a building in the hot tropical weather of Singapore.

In several other studies done, it is also proven that cool roofs absorb much less heat as compared with traditional dark-colour roofs as they are able to reflect the

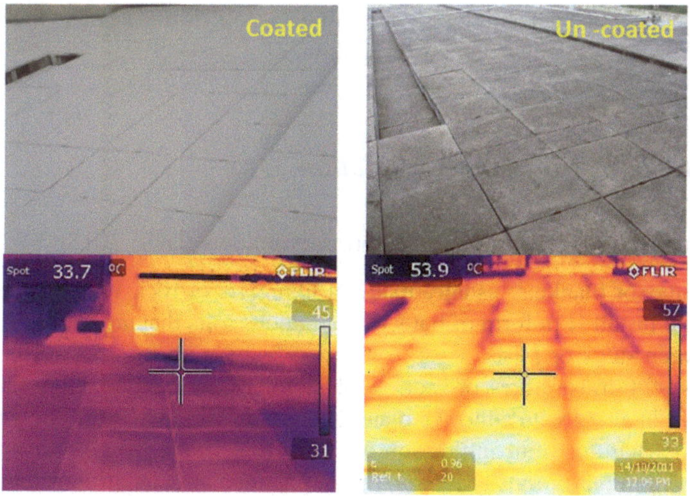


**Fig. 4.7** Thermal imaging study in Singapore showing that cool coatings can reduce the surface temperature by more than 20 °C (*Source* ERI@N, Singapore 2013)

sunlight back into the sky (Urban and Roth 2010). The effectiveness of cool roofs is measured by the following properties:

1. Solar reflectance: measures the ability to reflect sunlight and is the ratio of the sunlight reflected to the solar radiation that is absorbed. It is also known as albedo. A solar reflectance value of one represents total reflectivity (high albedo) and a value of 0 indicates that the roof surface absorbs all solar radiation.
2. Thermal emittance: measures the ability to emit or release absorbed heat and is the ratio of the heat emitted to the heat absorbed. It is usually expressed as a decimal fraction between 0 and 1, or as a percentage value.
3. Solar reflectance index (SRI): measures the overall ability to reject solar heat and incorporates both solar reflectance and emittance. A standard black surface has a SRI of zero and a standard white surface has a SRI of 100.

Cool roofing materials such as white thermoplastic membranes typically have a solar reflectance of 80 % and thermal emittance of at least 70 %. This is significantly higher than asphalt roofs, which are able to only reflect up to 26 % of solar radiation typically. Stainless steel roofs also have a very high solar reflectance index (SRI) of 100–115, while the SRI of a perfect mirror is approximately 122 (CRRC 2015). Any roof surface can also be made reflective by applying a solar reflective coating. These coatings are typically paint formulations that achieve a very high SRI and are sometimes referred to as 'cool paints'.

The Cool Roof Rating Council (CRRC) has created a rating system for measuring and reporting the solar reflectance and thermal emittance of roofing products. This system has been put into an online directory of more than 850 roofing products and is available on the web at: http://coolroofs.org/. It should be noted that although cool roofs are helpful in achieving cooling energy savings during the hot summer, it can erode the benefits of solar heat absorption during cold winters. Thus the overall benefits and annual net energy savings due to cool roofs need to be evaluated carefully for the particular climatic conditions.

### 4.5.3 High Performance Insulation

Insulation usually comes in different forms as follows (DOE 2015):

(a) Battings or blankets: usually made of fibreglass or rock wool (mineral fibre), can be installed in form of batts or continuous rolls that are stuffed into spaces between studs or joists.
(b) Loose-fill: loose fibres or rock wool, fibreglass or cellulose that are filled or blown in along with adhesive materials into building cavities. They can offer a firm fit and avoid air infiltration along with resistance to heat transfer.
(c) Foamed in place: mostly polyurethane, phenolic or cementitious materials that are sprayed directly into cavities where they are allowed to expand and fully seal the cavity. They offer a good air barrier and also have high R-values.

(d) Rigid board: plastic polyurethane or expanded/extruded polystyrene boards that are mechanically fitted on to the surfaces such as exterior walls, foundation and stem walls, concrete slabs and ceilings. They also provide structural strength while being low weight with high R-values to offer good thermal as well as acoustic insulation.
(e) Reflective films: radiant barriers that are mostly made out of aluminium foil with paper, cardboard or plastic backing that is integrated into housewrap or into rigid insulation boards. they are most effective in hot climates to reduce summer heat gain.
(f) Structural Insulated Panels (SIPs): prefabricated insulated structural elements that can be used in the building envelope system (walls, ceilings, floors, and roofs). They are made in a factory and shipped to job sites, where they are connected together during the building construction. They provide superior and uniform insulation compared to more traditional construction methods (stud or "stick frame"), offering energy savings of 12–14 % (DOE 2015).

#### 4.5.3.1 Sandwich Walls

Sandwich walls or Sandwich panel claddings typically consists of two aluminium skins with a polyurethane (PU) or mineral wool core forming a light-weight cladding panel with very good insulating properties (see Fig. 4.8 for illustration of a sandwich wall panel). In some cases, the metal cladding can be also made of light weight material through perforations that allow air-flow through. They offer external insulation and reduction in solar heat gain for opaque walls. They can also be integrated as architectural enhancement and can help the building skin weather against rain and excessive heat and light exposure.

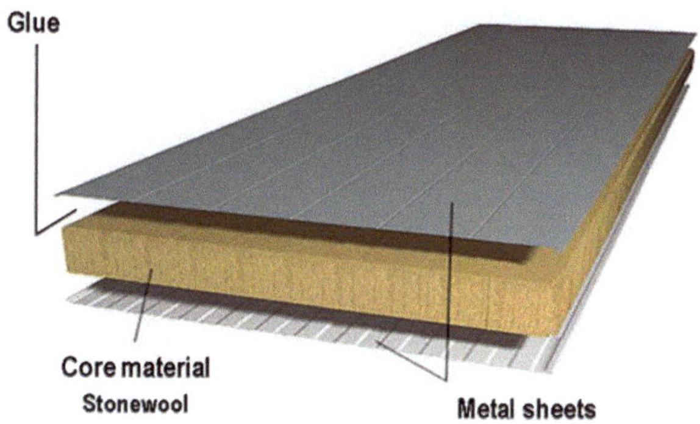


**Fig. 4.8** Illustration of a sandwich wall panel used in building insulation

#### 4.5.3.2 Thermal Insulation Plaster

Compare to conventional cement sand plaster, the thermal insulation plaster helps to cut down the heat transfer from the exterior to interior of walls and vice versa. It consists of materials that are mixed with cement, such as styrofoam beads, to enhance the insulation properties of cement, while being easy to apply. It can also save the use of sand in construction and provide a lightweight application that can enhance the fire resistance properties and recyclability. Figure 4.9 shows application of a thermal insulation plaster on the exterior walls of a building.

### *4.5.4 High Performance Glazing*

Windows and more typically glass windows (i.e. glazing) are important element of the building envelope and also offer significant energy saving opportunities. As described earlier in Sect. 4.4, the glazing energy performance can be determined by three main properties: insulating performance (U-value), solar heat gain coefficient (SHGC), and visible light transmittance (VLT or VT).

In addition to conventional single-pane and double-pane windows or glazing units, there are now newer technologies available that can significantly improve their performance on the above factors. These technologies are multiple pane glazing (triple pane, quadruple pane, etc.), inert gas fills, Low-E (low emittance) glass coatings, selective transmission films and adaptive glazing.

**Fig. 4.9** Photo showing application of a thermal insulation plaster (MagorTherm 2015)

#### 4.5.4.1 Multi-pane Windows and Gas Fills

The insulating performance (i.e. U-value) of glazing units can be improved significantly by reducing the convective flows within the unit by sub-dividing the air-space in between and adding more panes (see Fig. 4.10 for a simple illustration). These interior panes can often be simply thin films while the inner and outer side panes are structural glass. The spaces between window panes can be filled with gases that have the ability to insulate better than air. Gases such as argon, krypton, sulphur hexafluoride, and carbon dioxide are generally used for this purpose. These gases have much lower conductivity than air and can greatly reduce heat transfer by convective currents, resulting in a lower overall U-value for the glazing unit. As a resulting effect, the inner surface of the glass can be maintained at a temperature closer to that of the indoors, thus enhancing comfort and reducing cooling/heating loads.

#### 4.5.4.2 Low-E Coatings

Low-E (low emissivity) glass coatings are made from microscopically thin and transparent layers of metal or metallic oxide (such as silver oxide) that can be designed to reflect UV and infrared radiation while transmitting light in the visible

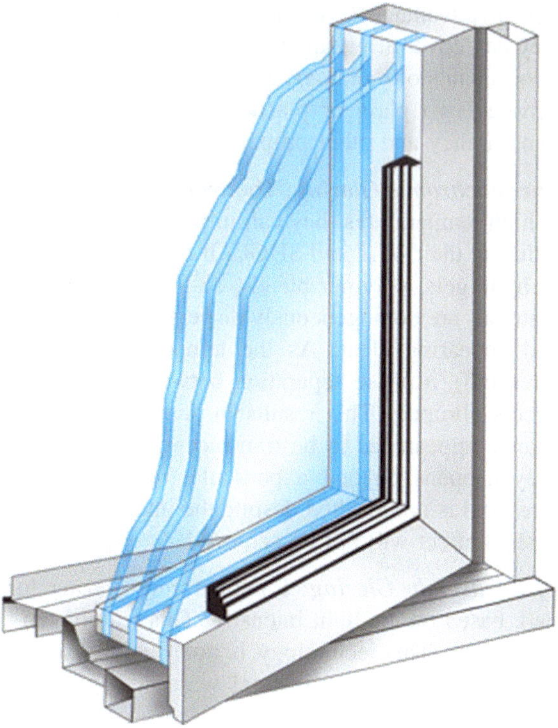

**Fig. 4.10** Simple illustration of a triple glazing window pane

wavelengths. Low-E coatings can be used in combination with tinted glass in order to avoid glare issues due to too much visible light transmission, while cutting significantly on solar heat gain.

#### 4.5.4.3 Selective Transmission Films

'Spectrally selective' window films selectively filter solar radiation by allowing visible light in while blocking heat (IR) and UV radiation. They are made of sophisticated combination of layered optical filters combined with nano particles to achieve the desired effect. These films appear mostly clear and are mostly effective in reflecting even far infrared radiations that are emitted by warm objects within the space. Thus these type of films can be effective in both summer (to cut long-wave infrared radiation from sunlight) and winter (to preserve heat from warm objects inside the space).

#### 4.5.4.4 Adaptive Glazing

Adaptive glazing systems have the ability to change their properties such as the visible light transmittance and solar heat gain coefficient, based on the outdoor conditions or on user on demand. These adaptive glazing technologies offer significant advantages over the static performance glazing systems described above as they can cover all weather conditions and any special requirements from the users or occupants of buildings. This is achieved by incorporating coatings and materials between two glass or plastic sheets that are capable of modulating the optical transparency and other properties in response to an external trigger.

***Thermochromic Glazing***: Thermochromic glazing are sensitive to temperature and at high temperatures they can turn from being clear to become dark or translucent, reducing their VLT and SHGC. It generally consists of thermotropic materials such as hydrogels, polymer blends, and block copolymers. At low temperatures, these materials are homogeneously dispersed in a matrix, thus minimizing light scattering and appearing clear. As the temperature is increased (beyond their switching threshold), a phase separation between the thermotropic domains and the matrix occurs abruptly. This results in scattering centres that reflect the light and they change appearance to be translucent. It should be noted that the phase switching only happens beyond a particular temperature threshold and sometimes if this threshold is not reached despite the intense solar radiation that is incident on it, the desired effect will not be achieved.

***Photochromic Glazing***: Photochromic glazing turn from being clear to become dark based on the light intensity. This feature can be commonly found in adaptive sunglasses. This technology is not yet fully scaled-up successfully at the level of using it for large commercial windows. As the switching is based on the light intensity, it might result in inappropriate action as the solar heat is not always proportional to the light that strikes the window, especially in case of low angled sun.

***Electrochromic Glazing***: Electrochromic glazing change their optical transparency and other properties by changing its colour or opacity when an electric field is applied. Once the field is reversed the optical properties are also restored back. In these type of windows, a thin, multi-layer assembly is sandwiched between the glass panels. The outside layers of the assembly are transparent electronic conductors that enclose a counter electrode layer and an electrochromic layer, with an ion conductor layer in between. When a voltage is applied across the conductors, the ions move from the counter-electrode to the electrochromic layer and cause the assembly to change colour. Reversing the voltage moves ions from the electrochromic layer back to the counter electrode layer, restoring the device to its previous clear state.

#### 4.5.4.5 Thermal Breaks

Although the U-value of the glass can be improved with the above technologies, it's also important that the window frame does not conduct heat around the glass. Typically window frames that hold the glass are made of aluminium or other light-weight metals. Ideally, the metal framing should be 'thermally broken' to separate interior metal elements from exterior elements. Thermal break is nothing but an insulating barrier between the inside and outside of the window frame. Thermal breaks can be made from materials such as high density polyurethane resins or fibreglass. Figure 4.11 shows an illustration of the use thermal breaks in triple glazed window frames.

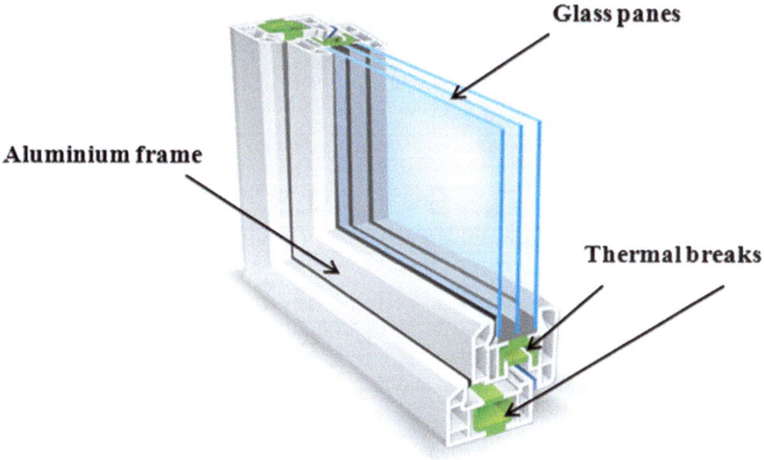


**Fig. 4.11** Illustration of a triple glazing with thermal breaks

The thermal break must be a poor conductor of heat in order to avoid heat from moving from outside to inside and vice versa. It also functions as a structural element in holding the two metal profiles together. Thermal breaks are also good for sound absorption and insulation compared to metal, which can also conduct sound easily.

#### 4.5.4.6 Overall Glazing Unit Performance

As seen in the above sections, the overall glazing energy performance not only depends on the glazing material and type of coating used, but also on the design of the whole window system, i.e. multiple glazing panes placement, frame, spacers, and the gas fill. To express the interaction of these components, the National Fenestration Rating Council (NFRC) has developed standards for rating the U-value (also known U-factor), SHGC, and VLT ratings of whole windows, including all of their components. Figure 4.12 shows an example of the NFRC label.

There are also software tools available that support the systematic evaluation of alternative fenestration systems. One such tool is COMFEN that is developed by the Lawrence Berkeley National Laboratory (LBNL). Similar to other modelling and simulation tools described in Chap. 3, this tool uses the Energy Plus simulation engine. The results from the simulations are presented in graphical and tabular format within the simplified user interface for comparative fenestration design cases to help users move towards optimal fenestration design choices for their project at hand.

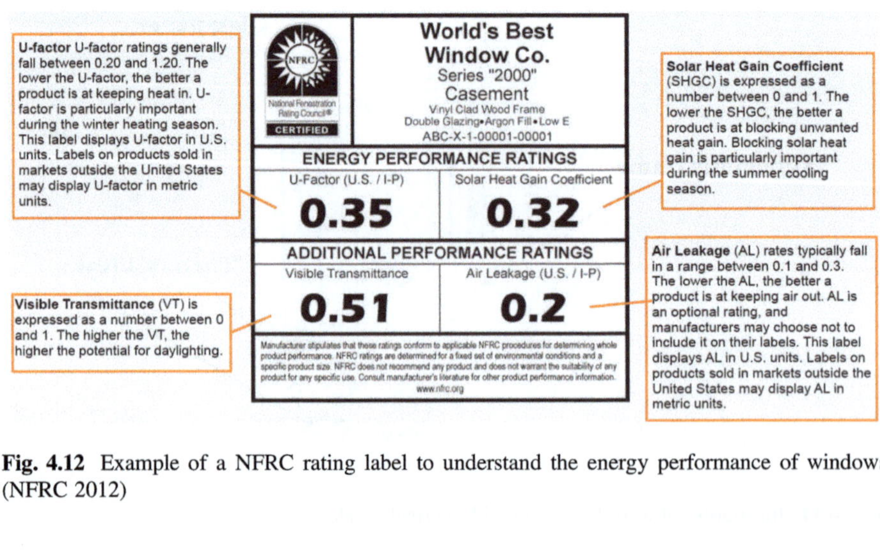


**Fig. 4.12** Example of a NFRC rating label to understand the energy performance of windows (NFRC 2012)

## 4.6 Passive Heating Technologies

Passive heating takes into account the energy of the sun to design for occupant comfort while reducing the energy used on mechanical systems for providing the same. Sunlight can heat a space via heat absorption in the solid walls or roof and also can enter the space through the fenestration, and heat interior surfaces. Some of the sun's light is long-wavelength infrared (IR) radiation, which is heat. In addition, the light of any wavelength absorbed by internal surfaces turns into heat in those materials. These materials then warm people in the room by conducting heat to them directly, by warming air which carries heat by convection, and by re-radiating their heat. This is all generally classified under the term 'solar heat gain'.

### *4.6.1 Massing and Orientation for Heating*

Massing and orientation are important design factors to consider for passive heating. The cold climates, the sun's heat is desirable for 'free' heating and reducing the use of mechanical or electric heaters. Hence, as a simple strategy the building surfaces exposed to the sun's path can be maximised to harness as much heat as possible from the solar heat gain. The opposite will be true for hot climates, where the undesirable solar heat gain can be avoided by orienting surfaces and fenestration away from the sun path and using massing strategies get shading effects to avoid direct exposure to the sun.

It is sometimes however not that straight forward to implement the simple strategy as described above site constraints, natural ventilation and day lighting also need to be considered for massing and orientation. The exposure time and hour of the day is also important. For example, heat gain on the east side can be acceptable or even useful, because it happens in the morning after the cooler night, whereas heat gain on the west side is not desirable at the end of a warm day.

### *4.6.2 Thermal Mass and Phase Change Materials*

Thermal mass in the building can store energy absorbed from the sun and release it slowly over time. Conversely, it can resist heating up too fast from solar radiation. It would be desirable to have building objects or materials with high thermal mass to absorb and retain heat, thus slowing the rate at which the space gets heated in the sun and cooled at night. In absence of any thermal mass, heat that has entered a space will re-radiate back and make the space overly hot when exposed to the sun. Also when the sun sets, the space can get overly cool when there is no thermal mass that has absorbed and retained the solar heat during the day.

In locations, with large differential temperature between the day and night (e.g. desert climate), the thermal mass can provide an effective mechanism to reduce mechanical heating and cooling loads. In locations that are constantly hot or cold throughout the day and night (e.g. warm tropical climates or sub-polar regions), the thermal mass may not be desirable as it can have a rather detrimental effect. This is due to the fact that all surfaces will tend to be at the average daily temperature of the surroundings, resulting in unwanted radiant gains or losses and hence occupant discomfort or huge energy penalties to counter the effect of thermal mass.

Thermal mass can be implemented in the building architecture as thick concrete floor slabs, water containers or tubes, and interior masonry walls with clay bricks or natural rock and stone. A large surface of interior thermal mass with direct exposure to sunlight is most beneficial for passive heating. Thermal mass can also be effective for cooling if the heat gain from the direct sun exposure can be effectively dissipated outside or to the ground without affecting interior spaces. However, such architectural implementations of thermal can lead to bulky buildings with excessive weight.

This problem of bulky thermal mass can be solved by using phase change materials to add thermal mass. These materials absorb and release thermal energy during the process of melting and freezing in form of latent heat. Common phase change materials such as wax and molten salts can be used for storing heat energy at appropriate operating temperatures. As a relatively large amount of heat is required for phase change at a given temperature, the phase change material acts as thermal mass that can absorb heat from hot surroundings and reject heat to cooler surroundings. The integration of phase-change materials in the building envelope can be achieved through micro-encapsulation and is a subject of ongoing research and development to achieve the right level of thermal mass effect at the right surrounding temperatures.

### *4.6.3 Trombe Walls for Passive Heating*

A Trombe wall, named after its French inventor Felix Trombe, is a system for indirect solar heat gain and incorporates the thermal mass and glazing greenhouse effects in order to achieve passive heating. It is essentially a wall with high thermal mass and preferably dark in colour that can absorb the solar heat when directly facing the sun. It is placed behind a glazing with a small air space in between. Figure 4.13 shows an illustration of this concept. The glazing further traps the solar radiation like a greenhouse and increases the heat that is absorbed and stored in the wall. This heat can then be conducted slowly inward through the masonry. By including upper and lower air vents in the wall, the air convection and flow in the space can be enhanced, as air heated in the Trombe wall flows into the room at the top and cooler air from the room enters the wall system at the bottom.

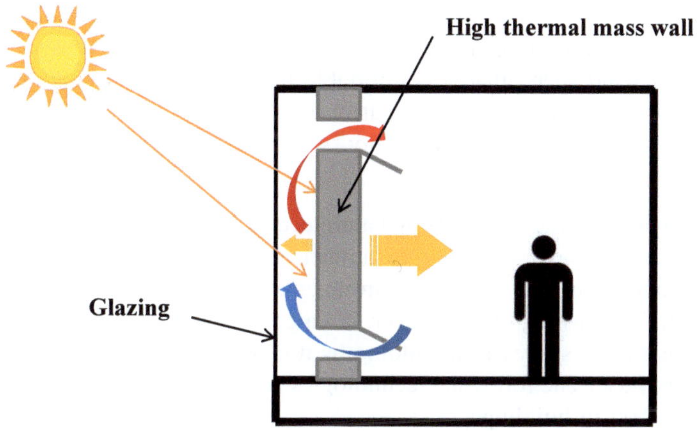


**Fig. 4.13** Trombe Wall for passive heating with interior vents

Trombe walls can function as highly effective passive heating system and also can enhance the silence and privacy in the room compared to rooms heated by forced-air systems. They also require little space to operate and can be easily integrated into the building architecture to enhance occupant comfort in cold climates through passive heating alone.

## 4.7 Passive Cooling Technologies

Passive cooling is useful in providing a cooler and more comfortable indoor conditions through the use of natural energy sources. The main method of passive cooling is through natural ventilation and use of ambient air for cooling. There are various strategies to induce natural air flow and use natural elements to enhance indoor thermal comfort.

### *4.7.1 Natural Ventilation and Cooling*

Natural ventilation uses natural outside air movement and pressure differences to passively cool and ventilate a building without the need for mechanical fans or air-conditioning system. For natural ventilation to be effective, it has to achieve the following:

1. Bring in enough volume of air to replace stale air in any space that may result in heat accumulation or accumulation of harmful gases/chemicals and odours.
2. Provide a wind speed that is comfortable for humans occupying the spaces or suitable for the activities carried out in the space.
3. Achieve temperatures and air quality, which are comfortable for the intended purpose of the space.

Often times, building designers and architects rely on natural ventilation only for spaces that do not have a continuous human occupancy or do not have needs for precisely controlling the indoor environment. Such spaces could be corridors, stairwells, toilets, atrium, etc. However, it's a good design practice to maximize the naturally ventilated spaces in a building as it can have a significant impact on its overall energy use. The following technologies could be used to enhance the natural ventilation effect in buildings.

#### 4.7.1.1 Massing and Orientation for Natural Ventilation

Massing and orientation are important aspects of design in order to effectively channel outside air through occupied spaces by playing with the building's height and depth. Tall buildings can significantly improve natural ventilation as wind speeds are faster at higher altitudes. At the same time, tall buildings if oriented correctly can also reduce the sun exposure.

Through good orientation and massing, buildings can maximize the benefits from cooling breezes in hot weather and block undesirable wind in cold weather. This is site and geography specific, but most weather data will have information on prevailing wind direction and intensity. The wind rose diagram as shown in Fig. 4.14 is a good graphic to refer to understand the wind conditions at a specific site.

The Beaufort scale is a good guide to understand the wind effects on people and the building planners and designers can use this to achieve various wind conditions in naturally ventilated spaces to achieve the desired effects. As can be seen from Table 4.1, the desirable wind speed for natural ventilation in buildings typically should be in the range 2–3 on the Beaufort scale.

As a massing strategy, providing void decks at the ground floor, higher floor-to-floor heights and void spaces in between buildings (as shown in Fig. 4.15) will improve air flow through and around buildings. This also helps to mitigate stagnant air flow areas. This however needs to be controlled to avoid excessive wind flow and/or wasting space unnecessarily.

#### 4.7.1.2 Openings for Cross Ventilation

While designing openings (doors and windows), architects and designers can plan for enhance natural ventilation by providing pathways for airflow through the structure. One common practice is to provide cross-ventilation by providing

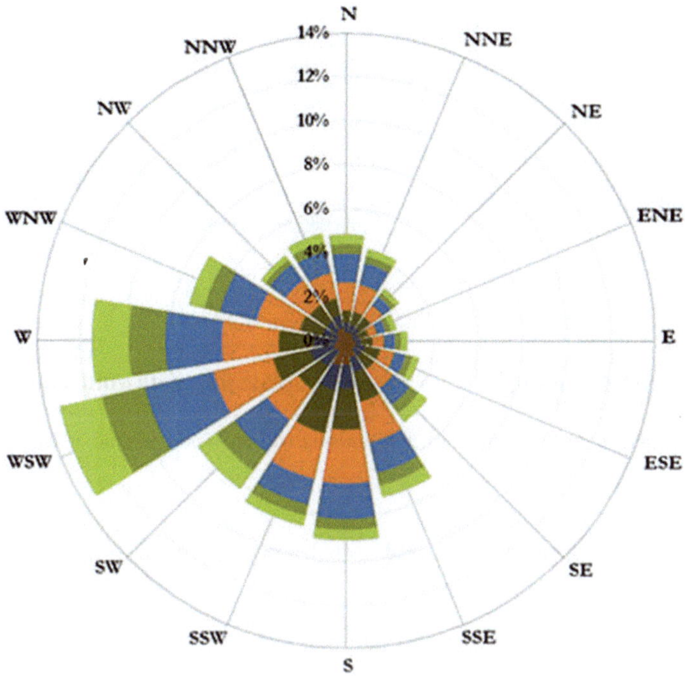


**Fig. 4.14** Wind-rose diagram showing wind speed and direction statistics for a particular region

**Table 4.1** Beaufort scale to understand effect of wind speed on human comfort (Penwarden and Wise 1975)

| Beaufort scale | Type of winds | Wind speed (m/s) | Effects |
|---|---|---|---|
| 1 | Calm, light air | 0–1.5 | Calm, no noticeable wind |
| 2 | Light breeze | 1.6–3.3 | Wind felt on face |
| 3 | Gentle breeze | 3.4–5.4 | Hair is disturbed, clothing flaps |
| 4 | Moderate breeze | 5.5–7.9 | Raises dust, dry soil and loose paper–hair disarranged |
| 5 | Fresh breeze | 8.0–10.7 | Force of wind felt on body |
| 6 | Strong breeze | 10.8–13.8 | Umbrella used with difficulty, hair blown straight, difficult to walk steadily, wind noise on ears unpleasant |
| 7 | Near gale | 13.9–17.1 | Inconvenience felt when walking |
| 8 | Gale | 17.2–20.7 | Generally, impedes progress, great difficulty with balance |
| 9 | Strong gale | 20.8–24.4 | People blown over by gust |

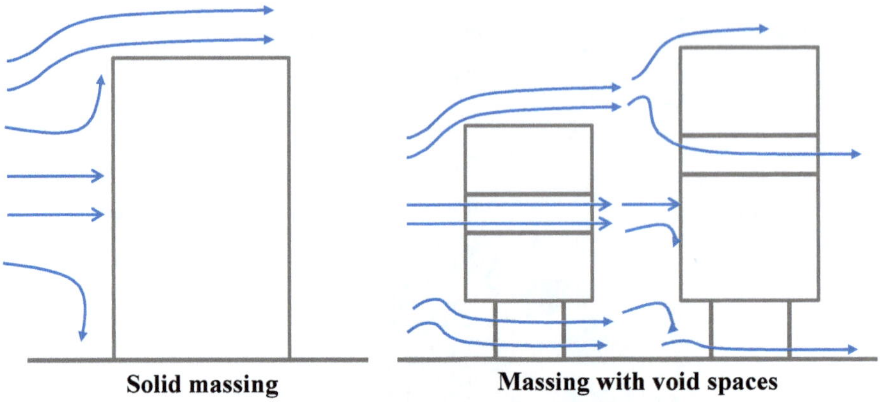


**Fig. 4.15** Enhancing airflow through better massing and provision of void areas

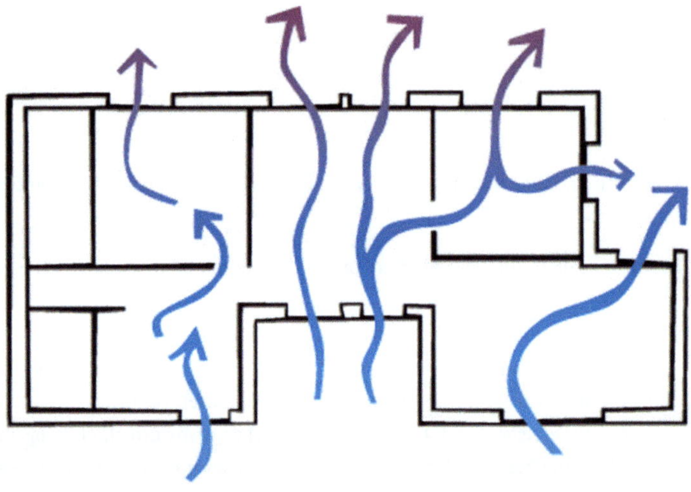

**Fig. 4.16** Openings for cross ventilation can enhance natural air circulation

windows or vents across the other side of the building, often directly opposite to each other (see Fig. 4.16). While designing openings for large spaces, it's important to factor in air-mixing and circulation within the space and placing opening directly opposite to each other may create dead-spaces or stagnant spaces where there are no openings. This could be mitigated by providing openings across from, but not directly opposite each other.

To effectively utilize the pressure differential for better wind speeds, inlets should be placed at high pressure zones and outlets at low pressure zones. Pairing a large outlet with a small inlet increases incoming wind speed. While designing openings for natural ventilation, other weather conditions such as rain has also to be considered and that can be done by providing louvers that can block rain while providing the wind flow opening.

#### 4.7.1.3 Wing Walls

Wing walls are architectural features that help in steering winds for enhancing the natural ventilation effect. They are built as vertical building envelope attachments that project outward next to a window or large opening in the building. Even a slight breeze against the wing wall can create a high pressure zone on one side and a low pressure zone on the other side. This pressure differential draws outdoor air in through an opening at one side of the wing wall where the pressure is high and out through the other where the pressure is low. Wing walls work well in sites with low outdoor air velocity and where the wind direction can vary considerably.

#### 4.7.1.4 Stack Ventilation

The stack effect is the passive movement of air through the building by leveraging on thermal buoyancy i.e. hot air rising due to its lower density. Stack ventilation makes use of this effect by incorporating openings in the building envelope at substantial height to allow the warm air to escape from the top. This results in negative pressure at the bottom that can draw-in cooler air from an opening near the bottom of the building. In tall buildings, this can be achieved by connecting the airflow of different floors and channelling it upwards through vents, towers and chimneys. Stack ventilation can be designed to be an effective mechanism for passive cooling and avoiding mechanical cooling and ventilation in summer. In winter however, the high temperature difference between the building interior and exterior can be problematic and result in over-ventilation and heat loss. When temperature differences between outside and inside are low, it may not be so effective and can result in under-ventilation, especially for the upper floors.

#### 4.7.1.5 Solar Chimney

Solar chimney works on a similar principle as the Trombe wall, but to enhance natural ventilation it uses the buoyancy and thermosiphon effect (stack ventilation). Solar heat gain warms a column of air trapped in between two walls, causing it to rise and pull new outside air through the building. In its simplest form, the solar chimney consists of a black-painted chimney. See Fig. 4.17 for illustration of this phenomenon.

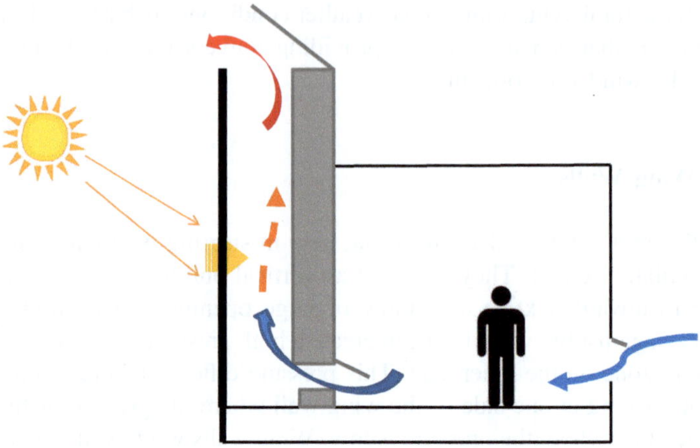

**Fig. 4.17** Illustration of the Solar Chimney with stack ventilation effect

To be effective for natural ventilation, the chimney head has to be higher than the roof level, and has to be constructed on the wall facing the sun directly. The suction created at the chimney's base can be used to ventilate and cool the building below. To further maximize the cooling effect, the incoming air may be led through cooler underground ducts before it enters the building.

### *4.7.2 Air Cooling in Dry Climates*

In dry climates, the principle of evaporative cooling can be put to use by passing incoming air over water features, vegetation or underground spaces that tend to humid and cooler. This can lead to cooler air being used for ventilation and have a good natural cooling effect. However, the mechanical energy required to push the air and overcome any pressure drop needs to be considered against the benefits achieved by routing the air for natural air cooling.

### *4.7.3 Double-Skin Facades*

The building façade is sometimes referred to as the skin of the building. A double-skin facade is structured in a way that there is an intermediate cavity in between the two skins where air can flow. The cavity can be ventilated by natural air circulation or through mechanical means. Figure 4.18 shows a photo of a building with double skin façade being assembled in the construction phase. The

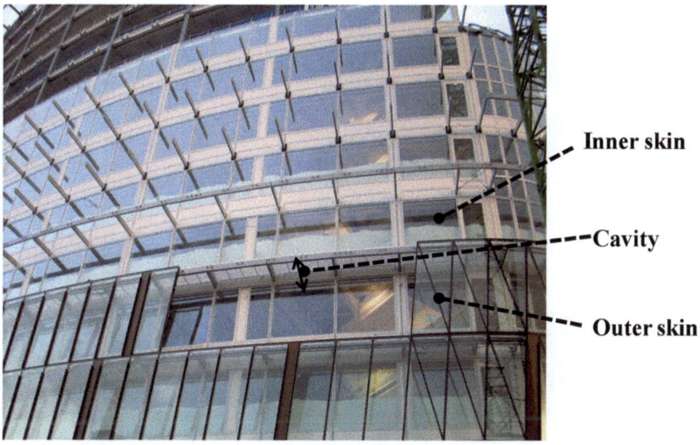


**Fig. 4.18** Photo of a building in London showing a double-skin facade being assembled

cavity can be used to vent out hot air to mitigate solar gain in hot climates and that can lead to reduction of cooling load. In cooler climates the solar gain within the cavity can be circulated to the interior space for heating.

## 4.8 Passive Lighting or Day Lighting

Passive lighting is nothing but using the daylight to illuminate the indoor spaces instead of overly relying on artificial lighting. This can significantly reduce the energy consumption in buildings, especially in spaces that are primarily used or operated in the day time. Daylight or natural light also has several physiological and psychological benefits. It is important however to distinguish daylight from direct sunlight, which can be too intense and result in excessively heating up spaces. What we refer to daylight here is the diffused natural light from the sky that is not too intense and can be used to lit up indoor spaces in buildings. In this section, several techniques to effectively use day lighting are described.

### *4.8.1 Daylight Apertures or Fenestration*

Openings or apertures in the building envelope (referred to as fenestration) can be used to introduce daylighting inside the building. There are different types of fenestration techniques that can used to introduce sufficient daylight in indoor spaces.

**Fig. 4.19** Photo showing daylight penetration through glass windows in building space

#### 4.8.1.1 Side Windows

Light coming from the side windows is the most common form of daylighting technique. However, the penetration of such light far into the building indoor spaces is limited. Hence, side windows as primary mode of daylighting work well only with shallow floor plans. Figure 4.19 shows a photo of a room with daylight penetration.

The orientation on side windows is also an important consideration. East or west facing windows are only effective at certain time of the day (i.e. morning or afternoon) and can also cause excessive glare or heating due to direct sunlight. The most even illumination without too much glare or heat can be provided by windows facing away from the sun's path. Other factors such as neighbouring buildings and tall features also need to be taken into account. Hence using side windows for daylighting should be carefully chosen for a particular location considering the sun path, massing and neighbouring developments.

#### 4.8.1.2 Skylight

Skylights are roof based openings that can be effective in bringing light deeper into the building, but can also cause direct sunlight issues in locations near the equator. The majority of commercial and industrial skylights are installed on flat roofs, where the skylight can have an almost unblocked view of the full hemisphere of the sky.

Skylights are available in a wide variety of sizes, shapes and forms (see Fig. 4.20) to match the building and structural requirements. They can range from simple rectangles to pyramids, ridges and other complex polygons. They can also be available in different sizes and be small, to fit between rafters, or large enough to cover the entire length of the building. Figure 4.21 shows photo of a large bubble-shaped skylight installed on spacer frames above an open walkway in a building.

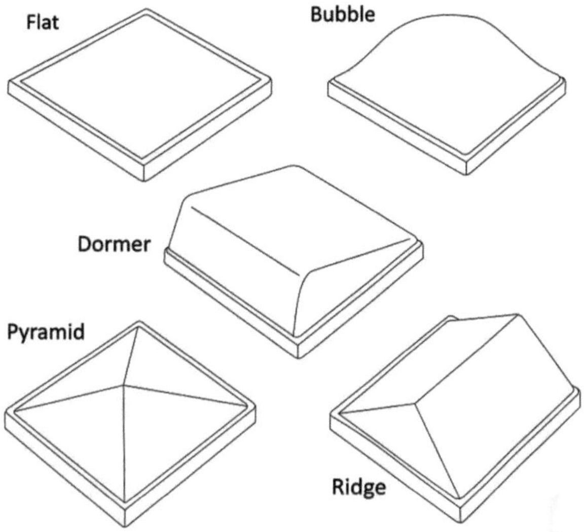


**Fig. 4.20** Different basic shapes of skylight

Apart from skylights, there also other kind of apertures to bring light in through roof. As shown in Fig. 4.22, these include clerestory, monitor, and saw-tooth. They differ from each other in how they bring-in the daylight into the space during different seasons and different times of the day.

### *4.8.2 Light Shelves*

Light shelves help light penetration into deep indoor spaces by bouncing visible light up towards the ceiling, which reflect it down deeper into the interior of a room. A light shelf is positioned above eye level and divides a window into a view area on the bottom and a daylighting area on the top (see Fig. 4.23 for illustration of the concept). The light shelf is usually a horizontal element that can be positioned externally, internally, or combined and can either be integral to the building, or mounted upon the building. It can be constructed of materials such as wood, metal, glass, plastic, or fabric. The structural strength, ease of maintenance, cost, and aesthetics of the building are the key considerations for material selection.

External light shelves can double up as shading devices as they can reduce the amount of incoming heat from direct sunlight apart from reflecting light for even distribution. However external light shelves are prone to accumulation of dust and even bird droppings and may require regular cleaning and maintenance. This is where internal light shelves have an advantage. Indoor light shelves however can take up indoor space and are not effective in shading from sunlight.

**Fig. 4.21** Photo of a skylight installed above an open walkway in a building

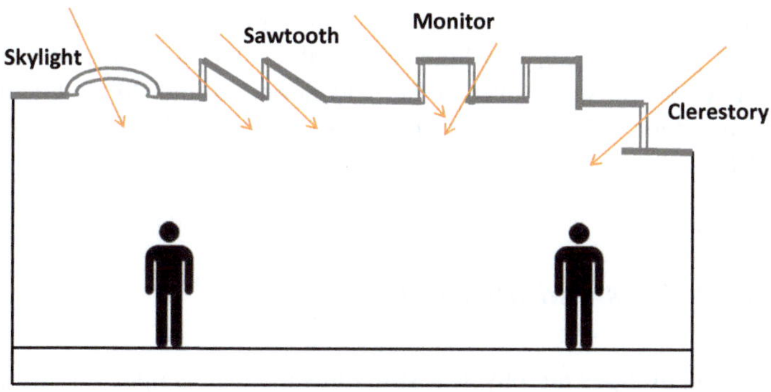


**Fig. 4.22** Different types of top (roof-based) lighting techniques

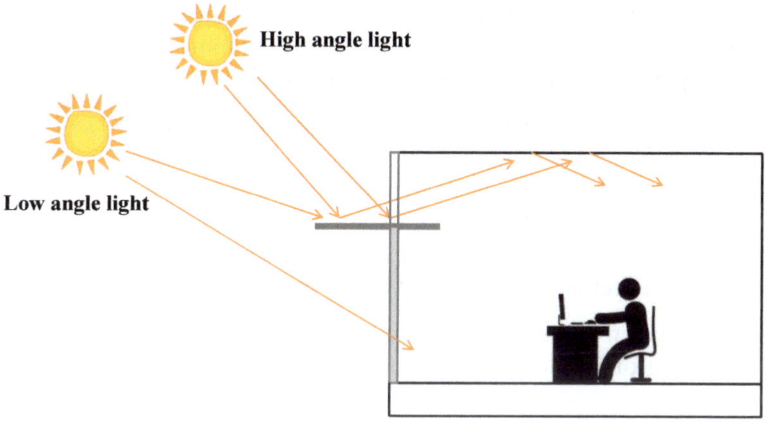


**Fig. 4.23** Illustration of the light shelf concept

### *4.8.3 Daylight Redirecting Glazing/Window Films*

Although light shelves are an effective method to reflect and bounce light inwards, they are additional architectural elements that require maintenance or take up space. There are now glazing materials or films available that are able to reflect light towards the ceiling and hence can possibly replace light shelves and be more cost-effective.

For example, the 3M Daylight Redirecting Film, utilizes micro-replication to redirect light that would have originally hit the floor a few feet from the window, up onto the ceiling, helping to light the room as deep as 40 feet from the window. The technology "micro-replication" refers to microscopic structures that are able to redirect as much as 80 % of light up onto the ceiling (3M 2016). Apart from 3M, there are other suppliers of daylight redirecting films such as SerraLux Inc. (SerraLux 2016).

Reflectance values from room surfaces will significantly impact daylight performance and should be kept as high as possible. It is desirable to keep ceiling reflectance over 80 %, walls over 50 %, and floors around 20 %. Of the various room surfaces, floor reflectance has the least impact on day lighting penetration (Ander 2014).

### *4.8.4 Light Pipes and Mirror Ducts*

Light tubes or light pipes are physical structures that act as optical wave guides for transporting or distributing light for the purpose of illumination. In their use for daylighting, they are often referred to as sun tubes or sun pipes. They function similar to a skylight, but are designed to optimally introduce light deeper into the building where traditional skylights and windows can't reach. Figure 4.24 shows the illustration of the light pipe concept for indoor lighting.

The light pipe consists of two essential components:

1. A light collection dome that is typically installed similar to skylight, but is hemispherical to bring light in from various angles.
2. A long hollow tube that is attached on one end to the light collection dome and the other end may look like a light fixture that introduces light into the desired space. This tube is typically made up of light reflecting metal and other surfaces that help to reflect and transport light over the length of the tube.

Another variation of the light pipe is a light duct or sometimes called as a mirror duct that consists of a hollow reflective duct to bring daylight deeper into the space.

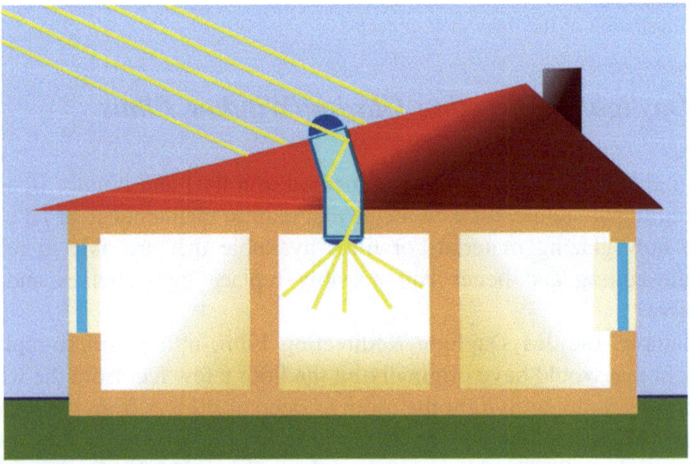

**Fig. 4.24** Light pipes for indoor lighting using daylight (Kuhn 2007)

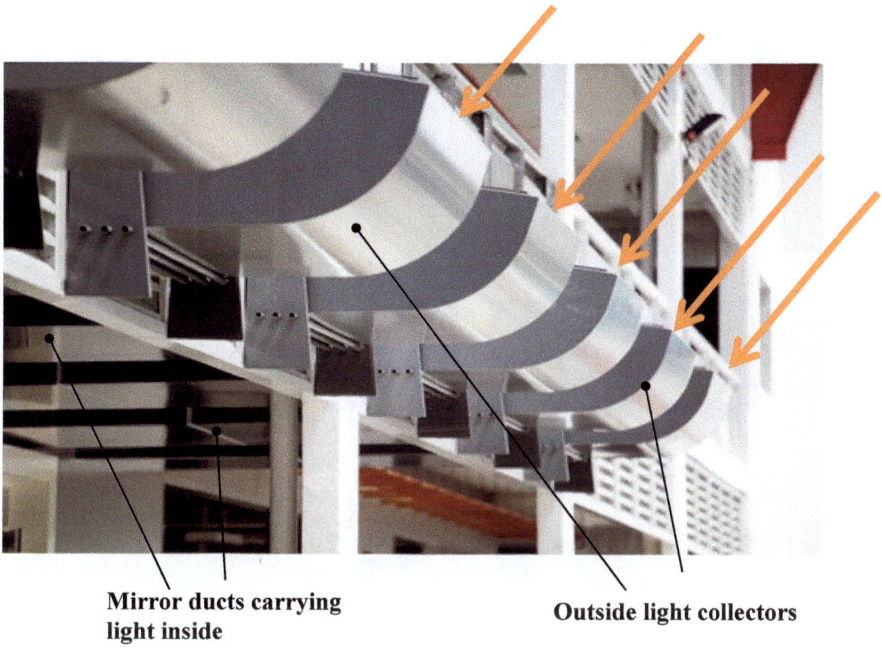


**Fig. 4.25** Photo of a mirror duct light collectors installed at a building in Singapore

While sun pipes are mostly installed vertically, mirror ducts are installed horizontally inside a space with light collectors placed outside as shown in the photo in Fig. 4.25.

Mirror ducts are designed to capture daylight through external collectors and the light is channelled into horizontal reflective ducts that are incorporated within the false ceiling. The light then exits through the ceiling apertures or illuminating units into the workspace below. The main advantage of the mirror ducts is that the light it brings into the space is usually glare free.

### *4.8.5 Transparent Insulation*

Transparent insulation differ from traditional building insulation in its ability to transmit daylight through. It typically consist of either glass or plastic material arranged in a honeycomb mesh, capillaries or closed cell construction. Alternatively, granular or monolithic silica aero gel can be used to achieve higher insulation values. The optical and thermal properties can be fine-tunes based on the material selection, its thickness and arrangement. Figure 4.26 shows some examples of transparent insulation materials. Generally transparent insulation is suitable more for cold climates rather than hot climates as it transmits solar radiation through or absorbs it as an thermal mass that can be used for passive heating.

**Fig. 4.26** Transparent insulation examples (Happold 2016)

### 4.8.6 *Measuring Daylight Effectiveness*

In case of day-lighting, the illuminance levels can vary throughout the day depending on the conditions of the sky. A measure of day-lighting effectiveness in a building space is the Daylight Factor (DF). Daylight factors are expressed as the fraction of natural light falling on a work surface compared to that which would have fallen on a completely unobstructed horizontal surface under same sky conditions. The calculation of DF is generally done using standard overcast sky conditions to represent the worst-case scenario for design. The ideal DF for indoor activities is in the range of 2–5 % as anything below that is considered poorly lit and anything more than that can cause implications on thermal comfort. The percentage of working hours when lighting needs are met by day-lighting alone is termed at the Daylight Autonomy (DA).

### 4.8.7 *Integration with Electric Lighting Controls*

Often times, daylight cannot be fully reliable to achieve continuously uniform illumination for visual comfort. Hence it has to be well controlled and augmented by integration with artificial or electrical lighting systems that will be discussed later in Chap. 5. An integrated control to enable electric light activation on demand can achieve significant energy savings while leveraging on daylight benefits to the maximum. These controls can be configured as follows:

1. Switching controls: on-and-off controls that turn the electric lights off when there is sufficient daylight available.
2. Stepped controls: control individual lamps within an electric luminary system to provide sufficient artificial light to augment the daylight as required.
3. Dimming controls: continuously adjust electric lighting by modulating the power input to lamps to complement the illumination level provided by daylight.

The main element of the daylighting based control is a photo sensor that detects or measures the intensity of daylight and can give appropriate digital signals to the controller for switching or dimming.

## References


3M, Daylight redirecting film. Available Online at: http://solutions.3m.com/wps/portal/3M/en_US/Window_Film/Solutions/Markets-Products/Government/Daylight_Redirecting_Film/. Accessed 15 June 2015

Ander GD (2014) Daylighting. Available online at: http://www.wbdg.org/resources/daylighting.php. Accessed 20 Sept 2015

Autodesk (2015) Available online at: http://sustainabilityworkshop.autodesk.com/building-design. Accessed 12 Aug 2015

CRRC (Cool Roof Rating Council) (2015) Available online at: http://coolroofs.org/. Accessed 15 Sept 2015

DOE (Department of Energy) (2015) Energy saver—types of insulation. Available online at: http://energy.gov/energysaver/types-insulation. Accessed 20 Sept 2015

ERI@N (Energy Research Institute at the Nanyang) (2013) Technological University (NTU), Singapore

Happold B (2016) Transparent insulation, updated 22 March 2016. Available online at: http://www.designingbuildings.co.uk/wiki/Transparent_insulation. Accessed 25 Mar 2016

Kuhn S (2007) Wikimedia commons. Available Online at: https://upload.wikimedia.org/wikipedia/commons/thumb/b/bf/Sonnenrohr.svg/1200px-Sonnenrohr.svg.png. Retrieved 12 Nov 2015

Levolux (2012) The cold facts about hot offices, 22 Feb 2012. Available online at: http://levolux.blogspot.sg/2012_02_01_archive.html. Accessed 20 Feb 2016

Magortherm (2015) http://www.magortherm.com/

NFRC (National Fenestration Rating Council) (2012) Available online at: http://www.nfrc.org/WindowRatings/. Accessed: 21 Feb 2016

Penwarden AD, Wise AFE (1975) Wind environment around buildings. BRE report, London

Sefaira (2013) http://sefaira.com/resources/3-ways-to-win-more-bids/

SerraLux (2016) http://serraluxinc.com/serraglaze-daylight-redirecting-film/serraglaze-behavior/

Urban B, Roth K (2010) Guidelines for selecting cool roofs. U.S. Department of Energy. Available online at: https://heatisland.lbl.gov/sites/all/files/coolroofguide_0.pdf. Retrieved 3 Nov 2015

# Chapter 5
# Active Design Technologies

In the earlier chapter we looked at Passive Design strategies and technologies that are built into the building and passively help by working with the ambient conditions in improving reducing the overall carbon footprint of the building. They do not consume any energy as such, but can help in reducing the operating energy use in the building. Although passive design technologies are an important aspect of designing green and smart buildings, they are not actively controllable and hence may not be responsive enough to tackle the changing demands and requirements in a building. Once designed in, the passive design features cannot be varied much to meet the specific operating needs.

Here is where we start looking at the active design technologies that can be dynamically controlled and hence can meet all the different operating conditions and requirements. Active design technologies consume energy and hence the most important aspect of such technologies for green buildings is their energy efficiency i.e. the ratio of the output delivered over energy input. We would be looking at active design technologies in the following areas:

1. Heating, Cooling and Ventilation
2. Lighting
3. Building services equipment such as pumps, lifts, escalators, etc.
4. Plug loads or receptacle loads.

## 5.1 Heating, Cooling and Ventilation

In buildings, occupant comfort is achieved in a controlled way through what is commonly referred to as Heating, Ventilation and Air Conditioning (HVAC) equipment. However, in warm climates there is much more requirement for cooling than heating. Also in tropical and humid weathers, its required to dehumidify the air to a comfortable level.



N.Y. Jadhav, *Green and Smart Buildings*, Green Energy and Technology,
DOI 10.1007/978-981-10-1002-6_5

Human comfort inside the building is however subjective and depends on various factors. To focus on thermal comfort of occupants in buildings, there are several that need to be considered:

1. Temperature of the air surrounding the occupant.
2. Relative humidity or water vapor/moisture in the air.
3. Air velocity or the rate of air movement.
4. Radiant temperature from the surfaces surrounding the occupant.
5. Clothing insulation provided by the clothes the person is wearing.
6. Metabolism or the energy generated from the human body.

While the personal factors such as clothing insulation and metabolism are difficult to control by the building engineers, they affect the occupants' overall perception of thermal comfort. The American Society of Heating, Refrigerating and Air-Conditioning Engineers (ASHRAE) has developed software programs for calculating thermal comfort parameters and making predictions using several existing thermal comfort models. It's called the ASHRAE Thermal comfort tool and is available for purchase on their website. A free tool is available at: http://comfort.cbe.berkeley.edu/.

The purpose of the active heating, cooling and mechanical ventilation technologies should be to achieve optimal thermal comfort at the lowest possible energy expense. In this section we look at several technologies for active heating, cooling and mechanical ventilation.

### 5.1.1 Air Heating and Cooling Technologies

The most common method of providing indoor thermal comfort is by heating or cooling air that is forced through the spaces with mechanical fans. In this respect there is a heat/cool generation equipment and a heat transfer equipment that transfer the heat from the generation source to the air that is pumped in the space. Figure 5.1 shows a simple diagram of typical air heating and cooling system for buildings.

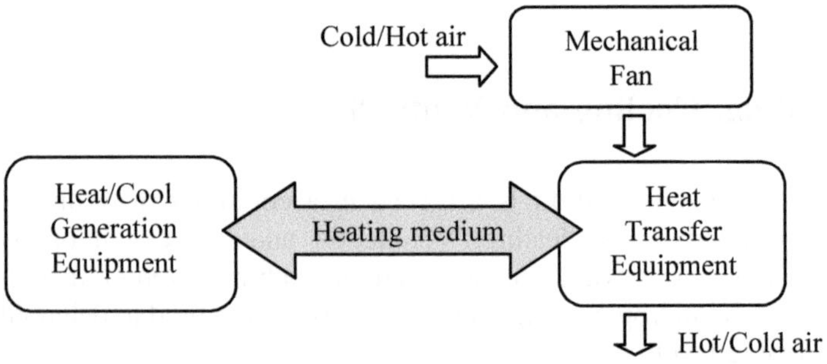


**Fig. 5.1** Simple representation of air heating and cooling system in buildings

#### 5.1.1.1 Furnaces and Boilers

Furnaces and boilers rely on burning fuel to create heat. The typical fuel used for building applications is natural gas or oil and less commonly coal, wood chips or other combustible fuels. Furnaces are used to directly heat the air, typically in smaller buildings. In case of boilers, water is heated in a closed loop and circulated around the building as a heating medium (hot water or steam) to heat the air.

The environmental factors considered for green buildings, when it comes to furnaces and boilers are:

1. Reducing harmful emissions from fuel combustion
2. Improving their fuel efficiency in order to reduce the amount of fuel used per unit of heat production.

The burning of fossil fuels results in emission of harmful gases as well as produces greenhouse gases that are responsible for climate change or global warming. Hence it's important to choose fuels with lower emission factors. Natural gas has the lowest emission factors amongst the commonly used fossil fuels. However, the choice of fuel depends on various other factors such as costs and local availability.

The fuel efficiency comprises of the fuel combustion efficiency, the stack heat loss and other heat losses from the outside surface of the boiler and losses during transmission. The combustion efficiency refers to the effectiveness of the burner in providing the optimum fuel-to-air ratio for complete combustion. The following measures can be used to improve overall boiler efficiency:

(a) Tuning up the operations of existing boilers using the best air-fuel ratio and cleaning tubes and other equipment to improve heat transfer. Typically, an analysis of the flue gas composition and temperature can provide insights into the boiler operating efficiency. As a rule of thumb, every 50 °C reduction in flue gas temperature can cause a 2.5 % increase in the boiler efficiency. However, there is a lower limit on the flue gas temperature as it might result in corrosion of the flue gas stack (200 °C for fuel oil and 105 °C for natural gas fired boilers).
(b) Increasing heat transfer in fire tubes by installing turbulators to create more turbulence and better mixing.
(c) Insulating the jacket and adjoining pipe connections to reduce heat losses.
(d) Installing soot blowers to remove boiler tube deposits that impede heat transfer between the hot combustion gas and water.
(e) Using economizers to recover heat from the stack flue gases and transfer it to het the combustion air or pre-heat feed water to the boiler. As a rule of thumb, for every 5 °C rise in boiler feed water temperature, the boiler efficiency improves by 1 %.

Most commercially available boilers are able to achieve combustion efficiency of well over 85 %, with the best in class traditional boilers being able to achieve up to

95 % combustion efficiency. Recent innovations in boiler technology such as 'pulse-combustion' boilers are able to achieve between 95 and 99 % combustion efficiency.

Pulse-combustion boilers operate similar to the internal combustion engines used in automotives. During their operation, pulse-combustion boilers extract the latent heat from the products of combustion by condensing the flue gas, which reduces its water vapour content and hence avoid stack corrosion problems even at lower flue gas temperatures. Pulse combustion boilers are hence highly efficient and achieve relatively very low exhaust flue gas temperatures of about 50 °C.

Also boilers that operate at lower loads that the maximum capacity are much less efficient than those operating at full capacity. Also in some cases the boilers are operated in on/off cycling mode to avoid operating at low part-loads. The on/off cycling mode is an inefficient way of operating boilers as there are efficiency losses due to cooler feed water and higher flue gas temperatures. An effective way to avoid the on/off cycling mode is to install modular boilers or multiple sized boilers that cater to varying loads. Modular boilers are highly recommended for application with largely varying loads such as schools, hotels and high-rise buildings as they can greatly improve the overall seasonal efficiency of the heating system.

#### 5.1.1.2 Electric Heaters

Electrical heaters or electrical resistance heaters use electrical current over large resistors to generate heat. The electrical resistors are organized as coils, which are then put in chambers where air or water can pass over them and get heated up (as shown in Fig. 5.2). Thus electrical heaters can be used as furnaces or boilers.

Electric resistance heating is 100 % energy efficient in the sense that all the incoming electric energy is converted to heat. However, most electricity is produced from coal, gas, or oil generators that convert only about 30 % of the fuel's energy into electricity. Because of electricity generation and transmission losses, electric heat is often more expensive than heat produced in combustion furnaces and boilers. There are however no local fumes emitted and if the electrical energy source is cleaner e.g. solar photovoltaics, electrical heaters are better alternatives for green buildings.

#### 5.1.1.3 Compression Chillers

Vapour-compression chillers can be driven by electricity, turbines or reciprocating engines. The electric driven (centrifugal, reciprocating or screw compressor) chillers are the most common in central chilled water applications. A typical compression chiller system consists of a compressor, a condenser, an expansion device, an evaporator, and other auxiliary equipment such as pumps. Figure 5.3 shows a simple schematic of a vapour compression cycle chiller system.

Fig. 5.2 Electrical heater resistor coils used in a electric boiler (Kristoferb 2016)

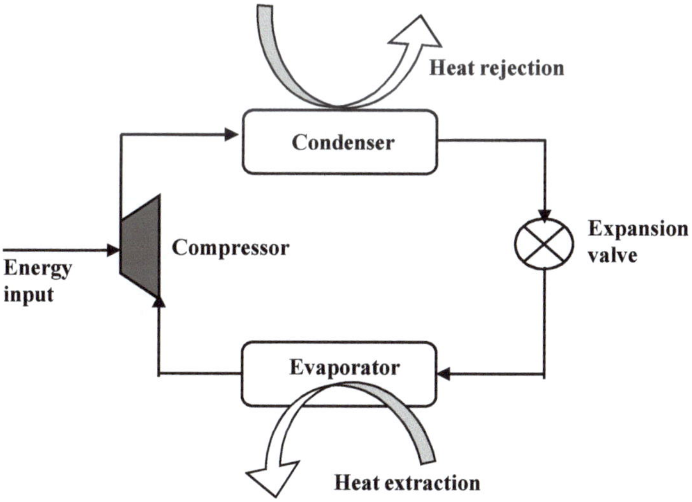


Fig. 5.3 Schematic of a vapour compression cycle chiller system

In this continuous cycle, the compressor runs with an external energy input to compress the refrigerant that expands at the expansion valve. The heat extraction from the expanded refrigerant vapours occurs at the evaporator and heat rejection of the compressed refrigerant is done by the condenser. Both the evaporators and condensers are mere heat exchangers. The evaporator is where the chilling occurs when water cools down as the refrigerant extracts its heat out. At the condenser, heat is extracted from the refrigerant and rejected to the ambient air (for air-cooled condensers) or water (for water cooled condensers that are typically connected to cooling towers).

The compression chillers can be used as stand-alone central chillers with the chilled water being circulated to several air handling units (AHUs) that provide the space cooling. Often times, they are also available as packaged units as follows:

(a) Split packaged units that have an air-cooled condenser and compressor installed outdoors and the refrigerant is circulated to the evaporator that is installed in an indoor air handling unit or sometimes referred to as fan coil unit (FCU).
(b) Vertical packaged systems that are designed for indoor installation.
(c) Rooftop systems that are typically located on the roof and also are integrated with a built-in heating system to provide both cooling and heating.

In case of compression chillers, there are two main considerations for green buildings:

1. The environmental impacts of the refrigerant used; and
2. The efficiency of the chiller in producing chilled water per unit of electrical energy input.

Most manufacturers of compression chillers use chlorofluorocarbons (CFCs) and hydro chlorofluorocarbons (HCFCs) as refrigerants. Several international agreements and federal regulations have placed some restrictions on the use of CFCs and HCFCs as refrigerants due to their high potential in ozone depletion and global warming. Due to this reasons, hydro fluorocarbons (HFCs) are now being used as they have no harmful effect on the environment.

The energy efficiency of a cooling system is popularly characterized by the term 'Coefficient of Performance (COP)'. The COP is defined as the ratio of the heat extracted for cooling divided by the energy input required. The maximum theoretical value for COP can be estimated using the ideal Carnot cycle COP. The Carnot cycle is an ideal that consist of isentropic (same entropy) compression and expansion and isothermal (same temperature) evaporation and condensation. In real systems, the ideal energy efficiency of a Carnot cycle cannot be achieved due to pressure losses in the lines and heat losses in the exchangers. The best commercially available centrifugal water chillers have a COP of about 7.0, which is about 70 % of the ideal Carnot cycle efficiency.

The COP is dimensionless number and most manufacturers provide their COP of cooling systems in full load conditions. The capacity of cooling systems is expressed in kW (kilowatts) and is defined in terms of the maximum amount of heat that can be extracted for cooling. In many countries, manufacturers use the refrigeration tons to rate the capacity of the cooling systems and use the unit kW/ton to express their energy efficiency. The conversion from kW/ton and COP can be done as follows:

$$\mathrm{kW/ton} = 3.516/\mathrm{COP} \tag{5.1}$$

There are various technologies available to enhance the COP of the chiller plant system as summarized in Table 5.1.

#### 5.1.1.4 Absorption Chillers

Absorption chillers operate using a concentration-dilution cycle (as shown in Fig. 5.4) that works on the principle of absorbing heat at low temperature and rejecting heat at high temperature. The two most common absorption systems employed in commercially available absorption chillers are Lithium Bromide (LiBr)

**Table 5.1** Technologies to enhance performance of a chiller plant

| Technology/Technique | Description |
|---|---|
| Automatic condenser tube cleaning | Fouling and scaling in the condenser tubes for water cooled chillers will cause a rise in condenser approach temperature resulting in energy waste. Automatic tube cleaning system utilizes the difference in water pressure in the condenser water to circulate sponge or rubber balls to clean the heat exchanger tube every 30 min. This prevents fouling and scaling built up in the condenser tube, thus enhancing chiller efficiency |
| Variable Frequency Drive (VFD) for compressor motor | The chiller compressor motor is often run at full speed, with the corresponding output of the chiller plant throttled with vanes, valves or other mechanical devices. A VFD can be effectively used to electronically vary the speed of the compressor motor, matching its speed to the load |
| Water side economizers or 'free cooling' | During cool weather conditions, particularly at night, cooling loads can be served with chilled water produced by the cooling tower alone, entirely bypassing an energy intensive chiller. A heat exchanger is used to cool the building loop chilled water with the cooling tower water vat low wet bulb temperatures |
| Smart refrigerant distribution | Refrigerant maldistribution in evaporators lowers capacity and efficiency in vapor-compression systems. Smart refrigerant distributors sense and direct the proper amounts of refrigerant to each evaporator circuit maintaining optimum performance |

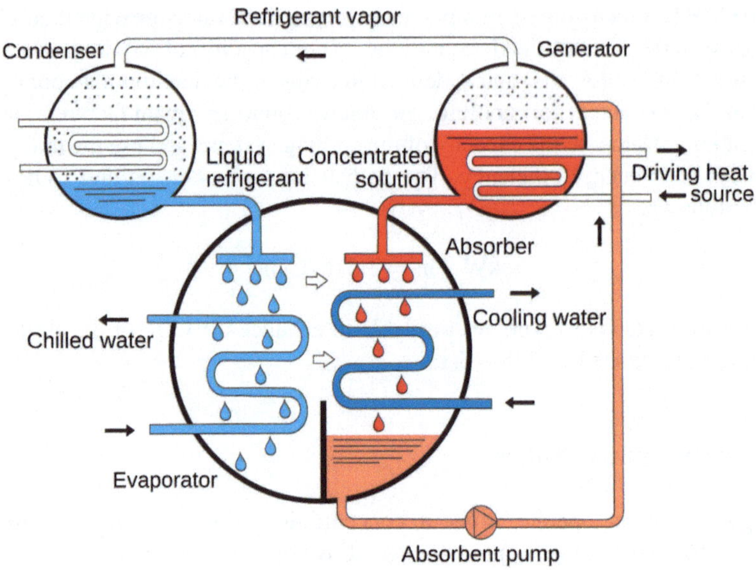


**Fig. 5.4** Schematic of the vapour absorption cycle

and water, and aqueous Ammonia solutions. In the case of the Lithium Bromide and water solution, the water is the refrigerant, while for the aqueous Ammonia solution, the Ammonia is the refrigerant.

The main difference between absorption chillers and compression chillers is that the absorption chillers use heat source to generate chilled water rather than electricity that is often used for compression chillers. Vapor compression cycle uses compressor, but the absorption chiller system uses a liquid pump to produce the pressure difference between condenser and evaporator. A liquid pump is much easier in work and cheaper than compressing a gas, so that it takes less work input. Also absorption chillers do not use refrigerants that can have harmful impacts on the environment as discussed in the earlier section.

Absorption chillers are typically much less efficient than compression chillers with COPs around 0.7–1.2. However, they can operate off of waste heat or solar heat while providing the cooling. In these cases, their efficiency is unimportant, because the energy source is free and does not cause any environmental impacts to use. They can also be used in places where electricity is unreliable or available at very high costs. Hence absorption chillers must be considered when there is 'free' heat source available for use and the overall benefits should be weighed against the lower cooling cycle efficiencies. There are also now multiple effect absorption chillers (double or triple effect) that have multiple condensers and multiple generators to allow for more refrigerant boil-off from the absorbent solution. Multiple effect absorption chillers can achieve COPs of close to 1.8, but are more expensive.

#### 5.1.1.5 Other Emerging Cooling Cycles

Apart from the vapour compression and absorption cooling cycles, there are other emerging technologies that are not yet popularly used for air conditioning systems. A brief description is provided below.

***Thermoelectric cooling***: Thermoelectric systems are solid-state systems that convert electrical energy into thermal energy (Goetzler et al. 2009). The 'Peltier effect' describes the unique behavior of thermoelectric materials. When a voltage is placed across a thermoelectric material (often a semiconductor), the voltage drives a current flow that moves electrons from one side of the material to another. These high energy electrons transport heat through the material. This creates a hot surface and a cold surface on opposite ends of the conductor and this temperature difference drives the system's cooling effect. Current commercially available thermoelectric systems offer low-efficiencies (COP of 1–1.5), but ongoing R&D in development of better materials may enable the creation of high-efficiency thermoelectric systems (up to COP of 6).

***Magnetic cooling***: Magnetic cooling is based on the magnetocaloric effect, a phenomenon in which a paramagnetic material exhibits reversible temperature change when exposed to a changing magnetic field (Dieckmann et al. 2007). A magnetic cooling system applies a magnetic field to a paramagnetic material, which aligns randomly oriented electron spins in the paramagnetic material, which is an exothermic process that raises the material's temperature. This heat is rejected from the material to its surroundings. Upon removal of the magnetic field, the magnetic spins return to their randomized state, which is an endothermic process that cools the material. The material then absorbs heat from the space to be cooled. During this step, the paramagnetic material returns to its original state. The heating-cooling cycle then starts again.

***Thermotunelling cooling***: Thermotunneling cooling is a solid-state system similar to thermoelectric systems but less mature and more complex. Thermotunneling systems rely on the transmission of electrons across a nanometer-length vacuum gap (driven by an applied voltage) to create a temperature difference between the two surfaces. Thermotunneling systems also make use of low work-function materials to further increase the efficiency of the system. The presence of the vacuum gap eliminates the backwards heat transfer issues that restrict the efficiency of conventional thermoelectric systems (EERE 2011).

### *5.1.2 District Heating and Cooling*

District heating and cooling systems are centralized heating/cooling plants that distribute the heating or cooling medium to multiple buildings. In effect the central heating and cooling services tend to be of a very large capacity and hence can arguably be operated more efficiently with central supervisory control instead of distributed controls. Also the synergies of heating and cooling systems can be exploited in an integrated way. e.g. Excess heat from furnaces and boilers can be used to drive absorption chillers and the heat rejected from chillers can be used to heat boiler feed water. Such systems can also use energy storage to buffer load patterns in buildings and possibly also combine heat and power (CHP) through co-generation of electricity and heat through gas/steam turbines.

In simple words, district heating and cooling systems produce steam, hot water and/or chilled water at a central plant and can also store the thermal energy. The steam, hot water and/or chilled water is then piped underground to individual buildings for space heating, domestic hot water heating and air conditioning. As a result, individual buildings served by a district energy system don't need their own boilers or furnaces, chillers or air conditioners.

The following are the benefits of district heating and cooling plant systems:

1. A combined heat and power plant can significantly increase (almost up to double) the fuel efficiency compared to a standalone furnace or thermal electricity generation unit.
2. The emission associated with heating and cooling systems such as burning of fossil fuels and use of refrigerants can be confined to a central location and hence can be controlled more stringently. This can improve the air quality as well as reduce the noise-levels that are typically associated with heating/cooling equipment.
3. Space optimization can be achieved with a much better effect in district heating/cooling systems as individual buildings need not cater for boiler and chiller plant rooms. This also gives the flexibility to architects to design versatile spaces for occupants and community as well as aesthetically pleasing buildings. No boilers or furnaces and roofs free of smoke stacks and cooling towers means substantially greater building design flexibility.
4. The central plants and proper space planning also enhances the ease of operation and maintenance as well as safety at the building level. The operation and maintenance can be planned and controlled centrally to achieve minimal disruption to building occupants and users.

However, in order for such central systems to be energy and cost-effective, the heating/cooling load density should be relatively high and preferably co-exist within the same district. The costs of the distributed piping system, thermal storage and central operation and control centre has to be carefully evaluated while considering the value of the above benefits.

### 5.1.3 Heat Pumps

In simple terms, heat pumps move heat from one place to the another instead of generating heat or cold. Heat pumps are designed to move thermal energy opposite to the direction of spontaneous heat flow by absorbing heat from a cold space and

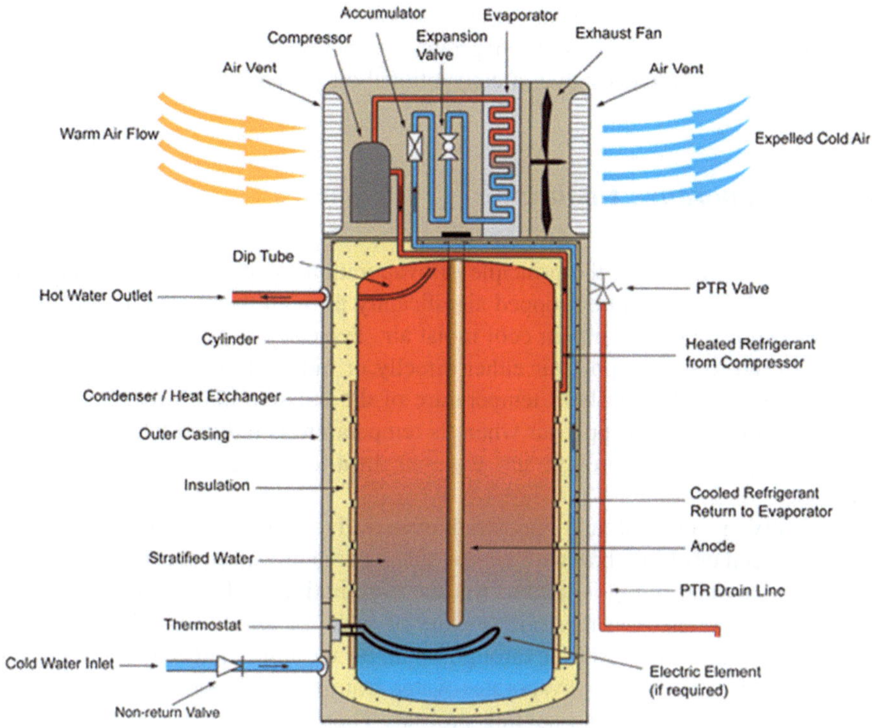


**Fig. 5.5** Integrated heat pump water heater with a wrap-around condenser/heat exchanger (DRET)

releasing it to a warmer one. Heat pumps work on the same principle of refrigerator or compression chillers (as illustrated in Fig. 5.3). Instead of the evaporator and condenser that are used in compression chillers, heat pumps extract or reject heat directly through tubings that are placed in the required zones or sometimes combined with heating/cooling coils or exchangers. Figure 5.5 shows an integrated heat pump for cooling and hot water generation.

Heat pumps can be used for both heating and cooling by reversing the refrigeration flow through the unit. The heat source or heat sink can be air (air-source heat pumps), water or ground (ground source or geothermal heat pumps). Ai-source heat pumps are typically used for heating buildings, but are less efficient in temperatures below 5 °C. Ground source heat pumps (GSHP) work by pumping a heat-conductive fluid such as water or propylene glycol through tubing in the ground. The digging and plumbing required to install a GSHP makes them very expensive and they are usually more efficient when the there is a significant temperature difference between the ground and ambient air. Water-source heat pumps typically use the adjoining water bodies (rivers and ponds) as heat sink. In warmer climates, air-source heat pumps can be used to heat water for hot water purposes while having a cooling effect on the air. Carbon dioxide ($CO_2$) heat pumps, which use carbon dioxide ($CO_2$) as the refrigerant, are relatively new technology that has promise of better efficiency than a conventional air-source heat pumps.

### 5.1.4 Evaporative Cooling

Evaporative cooling uses the principle of evaporation to cool air. The temperature of warm and dry air can be dropped significantly through the evaporation of water to water vapor, which results in cool moist air. Typically water sprays or a wetted media is used to cool supply air either directly or indirectly by allowing temperatures to approach the wet-bulb temperature of the ambient air. Direct evaporative cooling humidifies the supply air when its temperature is reduced whereas indirect cooling is performed through an air-to-air heat exchanger with no humidity addition.

Typically, residential and industrial evaporative coolers can be seen as an enclosed metal or plastic box with vented sides. Air is moved by a centrifugal fan or blower and a water pump is used to wet the evaporative cooling pads. The cooling units can be mounted on the roof or exterior walls or windows of buildings. To cool, the fan draws ambient air through vents on the unit's sides and through the damp pads. Heat in the air evaporates water from the pads which are constantly re-dampened to continue the cooling process. Then cooled, moist air is delivered into the building via a vent in the roof or wall.

Evaporative cooling principle is also often used as a supplementary unit in water cooled chiller plants. This unit is known as a cooling tower and is optimized to cool the water rather than the air. The cooling potential for evaporative cooling is dependent on the wet bulb depression, the difference between dry-bulb temperature and wet-bulb temperature. The energy efficiency of evaporative cooling is much better than other cooling methods in dry or arid climates. The only energy use is for fans and water pumps compared to compressors or other high energy equipment. The average COP of evaporative cooling can be in the range of 10–20 depending on the climatic conditions.

In arid and semi-arid climates, the scarcity of water makes water consumption a concern in cooling system design. However, as evaporative coolers use far less electricity and electricity generation itself usually requires a lot of water, the overall water usage can be comparable to water used in conventional chillers.

### 5.1.5 Dehumidification Technologies

Dehumidification is typically achieved in conventional air-conditioning units by cooling the air below its dew point and removing moisture by condensation. However, a more efficient way of achieving dehumidification is using desiccant materials (solid or liquid) that have a natural affinity for absorbing moisture from the air. As the desiccant removes moisture from the air, it releases heat and warms the air, which can then be cooled to desired comfort conditions using evaporative cooling or other cooling techniques described here.

The wet desiccant material (that absorbs the moisture from air) can be reused again by heating it and this process is termed as regeneration. Thus a typical desiccant dehumidification unit consist of a absorber (or conditioner) and a regenerator that work in a continuous cyclic manner to dehumidify air. In case of solid desiccants, this is achieved through a desiccant wheel that shuttles between the absorption and regeneration chambers. Liquid desiccants are typically pumped around between the absorber and regenerator. Figure 5.6 illustrates the concept of a desiccant wheel used for dehumidification of outside air.

In a liquid desiccant air-conditioning (LDAC) system, the heating at the regenerator can be achieved using either low-grade waste heat, water heated through solar means, or gas heating and heat pumps. When used with a free source of heating, these systems can achieve higher efficiencies than conventional technologies. Such systems can offer large energy savings in humid environments, especially when used in dedicated outdoor air systems (DOAS). The conventional compression chiller systems can be used in combination with LDAC and operate much more efficiently at higher chilled water temperature (boosting the COP) and can control the comfort conditions much better since dehumidification performance is no longer the driver for inferior temperature settings. LDAC can also be used in combination with radiant heating/cooling to achieve optimum thermal comfort at very low energy consumption.

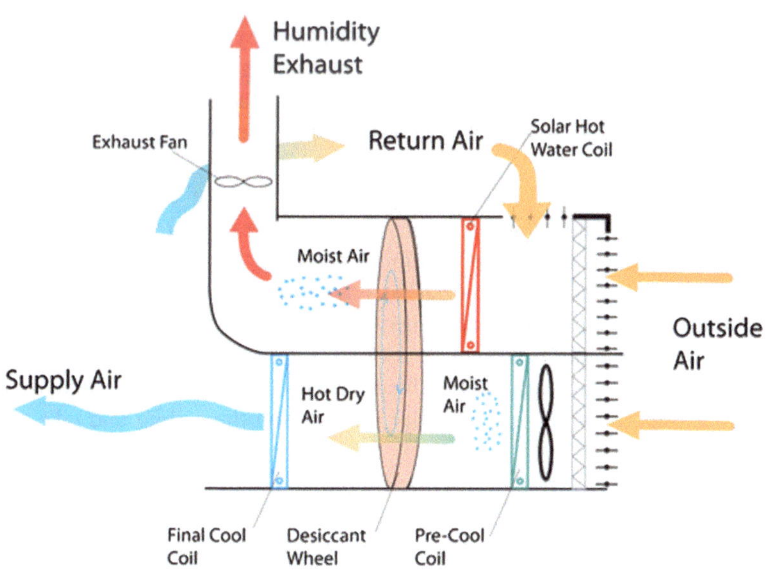


**Fig. 5.6** Desiccant Wheel working principle

Typically, silica gel is used as the solid desiccant and chemical solutions such as glycols, halide solutions, lithium bromide and lithium chloride are used as liquid desiccants. One of the main challenges with liquid desiccant technology is the entrainment of the liquid desiccant chemicals into air that can cause corrosion of air ducts and other health hazards. There is now research into membrane-based liquid desiccant units that use special membranes that ensure that the liquid desiccant is not in direct contact with the air and at the same time it allows for modular compact design of such system.

### *5.1.6 Radiant/Heating and Cooling*

Radiant heating or cooling rely on the phenomenon of radiation rather than convection through air. The heating or cooling is achieved through surfaces inside the space that radiate heat to the occupants. Such surfaces are typically embedded in the floor or ceiling or as separate radiant panels attached to the walls or ceiling. As radiant heating/cooling does not require forced air movement, a lot of energy savings can be achieved through such systems that can replace conventional heating or air-conditioning through air.

In case of radiant floors, the heating/cooling can be provided by electrical resistance heating or water (i.e. hydronic heating/cooling) pipes circulating through the floor panels. In some cases, the water pipes are embedded deep within the concrete floors to also achieve a large thermal mass that can also act as an energy storage or to dissipate heat for a longer duration throughout the day. Ceramic tile is the most common and effective floor covering for radiant floor heating, because it conducts heat well. Common floor coverings like vinyl and linoleum sheet goods, carpeting, or wood can also be used, but any covering that insulates the floor from the room will decrease the efficiency of the system. Radiant floor hydronic systems are the most popular and cost-effective radiant heating systems for heating-dominated climates.

In radiant ceiling systems, the hot/cold water can be circulated through the ceiling panels or chilled beams. Figure 5.7 shows the typical make-up of a chilled ceiling panel that consist of copper tubing enclosed in a metal panel that can be filled with high conductivity material such as carbon fiber or graphite. In case of cooling, the cold air sinks and warm air rises naturally to the chilled ceiling surface, thus also providing natural convective cooling (rather than forced circulation). They hence operate quieter and also help reduce the floor-to-floor height due to avoidance of air-circulation vents that are typically mounted on the ceiling and take up space inside false-ceilings.

Radiant systems are not useful for humidity control as they are not able to dehumidify the air as would be required in humid climate conditions. They are more suitable for dry climates or when combined with supplementary dehumidification systems such as those described in the earlier section on dehumidification technologies.

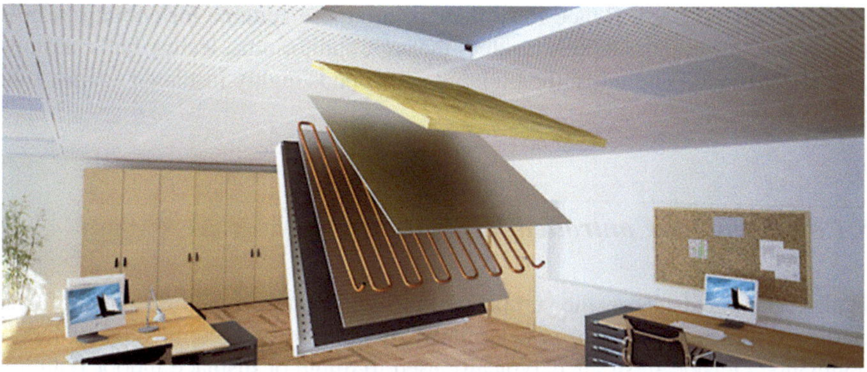

Fig. 5.7 Chilled ceiling panel components (SGL Carbon 2012)

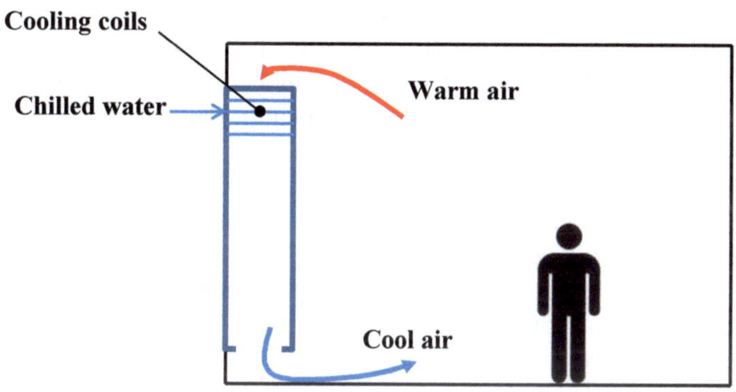


Fig. 5.8 Passive Displacement Ventilation illustration

### 5.1.7 Passive Displacement Ventilation

Passive Displacement Ventilation (PDV) is a room air distribution strategy where conditioned outdoor air is supplied at a low velocity from air supply diffusers located near floor level and extracted above the occupied zone, usually at ceiling height. Such systems drastically reduced the energy used for forced air circulation as done in the conventional ceiling mounted air diffusers. Passive Displacement Ventilation is typically used for cooling rather than for heating as natural convection through air stratification results in cooler air in the occupied zone (nearer to the floor than the ceiling). See Fig. 5.8 for a simple illustration of the PDV concept.

The main drawback of the PDV system is that the even air stratification can be hampered by obstruction such as furniture or desks and can result in 'dead spots' where the cooling is not effective at all. Also for perimeter regions which can get a lot of heat intake through the building envelope, the PDV system effectiveness will be drastically reduced. Hence PDV systems are more suitable for cooling in open spaces with fewer obstructions (such as theatres, auditoriums, lecture halls, atriums, gymnasiums, etc.) and when the building envelope is highly insulated against solar gain.

### 5.1.8 Demand Controlled Ventilation

In air-conditioned spaces, the ventilation rate or air change rate (the amount of fresh air introduced in the space) is determined by the rate of accumulation of indoor pollutants. Outside fresh air is typically used to dilute or displace indoor pollution in order to provide an acceptable indoor air quality. Mechanical ventilation uses fans to drive the flow of outside air into a building. This may be accomplished by pressurization (in the case of positively pressurized buildings), or by depressurization (in the case of exhaust ventilation systems). In kitchen ventilation systems, or for laboratory fume hoods, the design of effective effluent capture can be more important than the bulk amount of ventilation in a space for human comfort.

In any case, designers often assume a ventilation rate based on the occupancy schedule and certain assumptions on pollution rates within the space. Often times, the ventilation rate is over-estimated as the worst-case scenario is used to determine the ventilation rate. Demand controlled ventilation introduces dynamic ventilation rate controls that are based on the actual demand rather than gross estimates. For example, ventilation rates can be reduced during the many hours of operation when spaces are vacant or at lower than peak occupancy. When ventilation is reduced, building owners or operators save energy because it is not necessary to heat/cool or dehumidify as much outside air.

In case of office environment, demand controlled ventilation can be occupancy based or can also involve active measurement and control of carbon-dioxide, the main indoor air pollutant introduced by human breathing. For laboratories and spaces that require dilution of chemicals and fumes, such systems have to rely on sophisticated measurements of the pollutant concentration in order to ensure that there is sufficient dilution at any point of time.

### 5.1.9 Duct Sealing Technologies

Commercial buildings that generate cooling and heating usually deliver conditioned air to the building interior through a duct network. These duct networks contain thousands of field-assembled joints that are susceptible to air and heat leakage based on the quality of the materials and the installation (Conant et al. 2004).

Air leaks remove conditioned air from the system, requiring the cooling system to expend more energy producing additional conditioning to meet building loads. Leaks also reduce the air pressure within the duct system, resulting in slower moving air and mechanical fans may need to run longer or faster to compensate for this reduction. Building design engineers typically oversize their systems to compensate for duct leakage.

However, manually finding and plugging the leaks in a duct system is an expensive and time-intensive task. Aerosol duct sealing is an easier technique for sealing holes in the building ductwork. Aerosol duct sealing systems use an adhesive-aerosol spray. The spray is designed to accumulate when it comes in contact with holes in the ductwork, thereby sealing them. The ductwork system must be sealed and pressured prior to implementation. When implementing, an atomizer sprays the adhesive-aerosol spray into the ductwork as a solution suspended in air. Driven by an induced pressure differential inside the duct system, the adhesive-aerosol spray flows through any openings or cracks in the ductwork, accumulating at each spot and eventually sealing the hole (Hamilton et al. 2003).

According to Hamilton et al. Aerosol duct sealing can reduce system energy consumption by 4–9 %, by reducing the duct-leakage rate down between 2 and 3 % (from typical rates of 10–20 %) (Hamilton et al. 2003).

### *5.1.10 Fabric Ducts*

Fabric ducting is an alternative to traditional steel ducting and diffuser, where air is transported through ducts made of permeable fabric. They can significantly reduce the pressure drop and leak issues for conventional steel ducts, thus making them much more energy efficient. As fabric is easier to work with the installation and maintenance time is also drastically reduced. They also have fabric that can be made by environmentally friendly material and material that is washable. This can further enhance the indoor air quality and human comfort. Figure 5.9 shows a photo

**Fig. 5.9** Photo of a site with fabric duct installed (DuctSox.com)

of a site with fabric duct installed. The only disadvantage of this system of ducts is that they have to be exposed and cannot be concealed above the false-ceiling.

### 5.1.11 Ductless Jet Fans

Ductless jet fans are good for ventilation of large working spaces or areas such as car parks, halls, workshops and corridors. They have the characteristics of being able to induce a large amount of air from surrounding areas and long throw. They are primarily used to replace the extensive amount of ductwork that is required for providing mechanical ventilation in large open spaces. They are more energy efficient as well as space saving as they can completely avoid the static pressure losses and space requirements in ductwork. They can also allow for individual zone control and demand-based activation that can increase the overall energy savings achieved. Figure 5.10 shows the photo of a ductless jet fan installed in a car park building.

The use of Ductless Jet Fan mechanical ventilation system for fire safety and smoke evacuation must be accompanied by rigorous Computational Fluid Dynamics (CFD) simulations to substantiate that the level of safety and function provided by prescriptive designs are met.

### 5.1.12 High Volume Low Speed (HVLS) Ceiling Fans

The High Volume Low Speed (HVLS) Ceiling fans, provide a more energy efficient alternative to conventional ceiling fans, especially for large open areas such as school halls, sport halls, open air dining areas and areas with high ceiling. These

**Fig. 5.10** Photo of ductless jet fans installed in a car park of a building in Singapore

**Fig. 5.11** Photo of HVLS fan installed in an industrial workshop space (Kestrel Services)

fans can move huge columns of air efficiently at low speed, thus providing efficient and effective mechanical ventilation. They can replace multiple small ceiling fans in large areas and provide non-disruptive air movement and de-stratification. The main disadvantage of such fans is that they need a larger minimum clearance height or about 4.5 m of 15 ft. Figure 5.11 shows the photo of a HVLS fan installed at a large industrial workshop space.

### *5.1.13 Liquid Immersion Cooling for Data Centers*

The primary cooling load in data centers is server cooling. Unlike humans, servers do not need fresh ventilation air. Hence, the traditional approach to cool data centers with air conditioning is not a necessity in data centers. A novel cooling method known as liquid immersion cooling is based on cooling the servers through submersion in a dielectric liquid that is thermally conductive. With direct conduction and convection cooling, this method can lead to much higher efficiencies compared to conventional air conditioning based cooling. One of the simplest examples of liquid immersion cooling is taking a standard air-cooled computer's hardware and submerging it in mineral oil. Mineral, oil being non-conductive and non-capacitive, poses no threat to the electronics. Another method of liquid immersion cooling is based on a two-phase with a liquid that is a dielectric fluid with a lower boiling point. The liquid evaporates, condenses and drips back to the purpose-designed tank, which can have condensing coils or a indirect condensing surface.

**Fig. 5.12** Demonstration unit for liquid immersion cooling system at NTU (EcoCampus 2016)

These technologies are still in early stages of commercialization, but hold a great promise of significant reduction the data centre cooling energy. It can also significantly reduce the space requirement of data centers by being able to pack servers densely and reducing the space required for auxiliary cooling equipment. However, server maintenance and reliability issues may arise and need to be addressed in the system design. Such systems are currently under test-bedding stage of research and development. One such test-bed is at the Nanyang technological University (NTU) in Singapore under the EcoCampus RD&D (Research development and demonstration) program (EcoCampus 2016). Figure 5.12 shows the demonstration unit developed at NTU.

## 5.2 Lighting

Apart from thermal comfort, visual comfort is another important aspect of buildings. Apart from the day-lighting techniques that were explored in the earlier chapter, artificial or active lighting is needed in every building to provide optimum visual comfort throughout the whole day. The goal of a green buildings would be to provide optimal visual comfort at the lowest energy expense.

### 5.2.1 Lighting Performance Metrics

Before we go deeper into lighting technologies, it would be good to understand some basic lighting metrics. The effectiveness or the "brightness" of light can measured with the following parameters:

(a) Luminous flux (or luminous power): the amount of light coming from a light source in all directions. It is mainly measured by the unit 'lumens'.
(b) Luminous intensity: the amount of light from the source that travels in a certain direction. It is measured in 'candelas'.
(c) Illuminance: the amount of light falling on a surface. It is measured by the units 'lux' (lumen/$m^2$) or 'foot-candles' (lumen/$ft^2$). This is the measurement that is most often used by building standards and codes specify the minimum light levels for specific tasks and environments.
(d) Luminance: the amount of light reflected by a surface. It is measured in candelas per square meter (cd/$m^2$), or Nits (in imperial units). Luminance is what we perceive when looking at a scene, or when using a camera. It is not a good measure of the light quantity but rather it measures the effectiveness of the exterior/interior design and finishes in reflecting light appropriately. It is also a good measure of glare.

The luminous efficacy of the lighting source is measured by the amount of luminous flux (lumens) it can generate per unit of electrical power (watt) and is expressed in lumens/watt.

### 5.2.2 Light Source Technologies

Traditional incandescent bulbs use a lot of energy to produce light and have a low lifetime (750–2500 h). They are hence slowly being phased out from new manufacturing in most parts of the world. In an incandescent bulb, electric current runs through a wire filament and heats the filament until it starts to glow. Incandescent lamps are still used in some limited cases for accent and specialty lighting, where the warm color, controlled brightness, instant-on, and dimming capabilities of these sources is needed. There are now lighting technologies available that consume only a fraction of the energy that an incandescent light bulb would use. The most common energy-efficient lighting types include halogen incandescent, CFLs, HIDs and LEDs. Halogen bulbs are a type of incandescent that are slightly more efficient than standard incandescent but less efficient than most other alternatives that are discussed below.

#### 5.2.2.1 Compact Florescent Lamps (CFL)

CFLs use about 70 % less energy than incandescent bulbs and are more long lasting (about 10,000 h). A CFL produces between 50–70 lm/W, compared to the 10–19 lm/W for an incandescent bulb. CFLs emit light when an electric current causes an internal gas-filled (argon and mercury vapor) chamber to fill with ultraviolet (UV) light, which is then emitted as visible light through a special kind of coating on the tube. The phosphor coating on fluorescent bulbs gives them their distinctive white color. All fluorescent bulbs require a ballast, a component that regulates the current going through the lamp and also helps it to kick-start. Figure 5.13 shows the components of a CFL bulb.

In CFLs, the ballast is integrated into the bulb and these self-ballasted lamps are easy to screw-in as substitutes for most incandescent bulbs. Hence, CFLs are good option for retrofitting existing incandescent light bulbs with energy efficient lighting. However, the performance of screw-in lamps is usually not as good as the separate lamp and ballast combination. Mercury in CFL lamps leads to environmental issues when they eventually end-up in land-fills at the end of the life. Specifying a low mercury CFL product and then recycling that lamp at the end of its life offers the best environmental solution to disposal of CFLs.

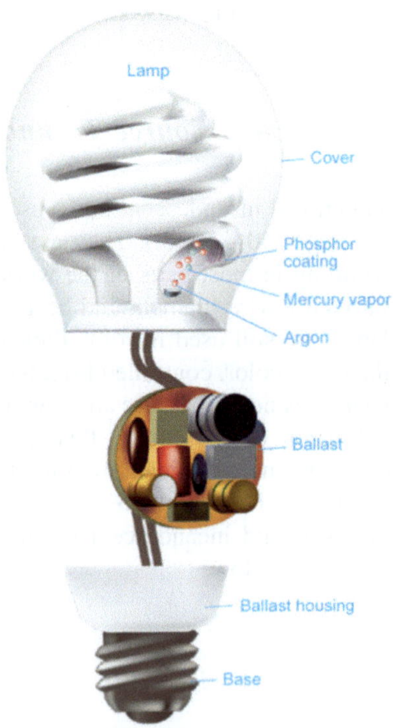


**Fig. 5.13** Components of a CFL bulb (Energy Star 2016)

#### 5.2.2.2 Linear Florescent Lamps (LFL)

Linear Fluorescent Lamps or fluorescent tubes have a separate ballast that is incorporated in the light fixture along with the tube. Ballasts come in two varieties:

- Magnetic ballasts, which are older and less efficient. They also produce a buzzing or humming noise in some fixtures.
- Electronic ballast, which are newer, quieter and much more efficient.

Efficiency upgrades for fluorescent tube lights require consideration of the ballasts because they contribute significantly to the overall energy draw of the fixture.

LFLs have a efficiency of 70–100 lm/W and a lifetime of 7000–24,000 h. LFLs are available in various tube diameters and typically classified based on their tube diameter with a T prefix on the label followed by a number that is a This measurement is expressed in eighths of an inch. For example, T8 tubes have a diameter of 1 inch. Typically, the narrower the lamp, the more efficient it will be. The older T12 LFL tubes (about 78 lm/W) are now been phased out and replaced with T8 (about 90 lm/W) and T5 tubes 9 about 99 lm/W) in most places around the world.

#### 5.2.2.3 Inductive Florescent Lamps

Inductive fluorescent lamps are white light sources with very good color rendering and color temperature properties. These lamps are energy efficient and offer extremely long life (over 100,000 h), good lumen maintenance characteristics, and instant-on capability (WBDG 2014). Induction lighting is essentially a fluorescent light without electrodes or filaments, the items that frequently cause other bulbs to burn out quickly. Thus, many induction lighting units have an extremely long life of up to 100,000 h. They are powered by a small generator (about the size of a fluorescent ballast) attached to the lamp via a short fixed-length cable.

The larger, diffuse nature of these sources makes them excellent for lighting larger volumes and surfaces. They are often used in place of low- to medium-wattage high intensity discharge sources because of the instant-on capability and reduced maintenance associated with the longer lamp life. This lamp source has promising application for indoor and outdoor lighting applications (WBDG 2014).

#### 5.2.2.4 High Intensity Discharge Lamps (HIDs)

As the name suggests, HIDs have a very high light intensity at a relatively good luminous efficacy of 50–150 lm/W. They are mostly used for outdoor lighting applications. There are different varieties of HID lamps now available:

- ***Low-pressure sodium lamps*** (60–150 lm/W) are among the most efficient available for outdoor use (street lighting, parking and security lighting), they are only useful for certain applications because of their long start-up time, cool-down time, and poor color rendition.
- ***High pressure sodium lamps*** (50–140 lm/W) that produce yellow light.
- ***Mercury vapor lamps*** (25–60 lm/W).
- ***Metal halide lamps*** (70–115 lm/W) that produce white light. Pulse initiated, or "pulse-start" metal halide lamps provide better color stability and longer life than previous technologies. PAR metal halide lamps with ceramic arc-tube enclosures are commonly used for accent lighting and highlighting in large spaces, and are now commonly used in retail applications.

#### 5.2.2.5 LED Lighting

In light-emitting diodes (LEDs), when electrical current passes through a semiconductor material, electrons and electron holes (atoms that lack an electron) combine, releasing energy in the form of light. LED lamps are the newest addition to the list of energy efficient light sources. LED fixtures use 75–80 % less electricity than incandescent bulbs, and can have a lifespan 25 times longer than incandescent light bulbs. They are now available in most markets even as replacements for standard screw-in light bulb fixtures (as shown in Fig. 5.14).

LEDs produce in the range of 27–150 lm/W, depending on the type of LED. LEDs are more durable than most other lighting alternatives and are more controllable because the light can be focused in a particular direction and the LED can be dimmed.

Through the use of LEDs and similar products, researchers are developing an array of lighting options that use solid objects (solid state lighting) rather than energy passed through a vacuum or gas to produce light. The useful life of LED lighting products is defined differently than that of other light sources, such as

**Fig. 5.14** LED lighting products available as replacement of standard screw-in bulbs (Landis 2016)

incandescent or CFL. This is because LEDs typically do not "burn out" or fail. Instead, they experience lumen depreciation, where the amount of light produced decreases and light color appearance can shift over time. LED product "lifetime" is set based on a prediction of when the light output decreases 30 % (Energy Star 2016).

LED chips need controlled direct current (DC) electrical power and an appropriate circuit is required to convert alternating current from the supply to the regulated low voltage direct current used by the LEDs. The heat produced from the power going into the product must be drawn away from the LEDs and this is achieved using a heat sink, which helps to dissipate the heat. The heat management is an important aspect of the LED reliability and long-term performance as its adversely affected by high temperatures. LED products use a variety of unique heat sink designs and configurations such as cooling fins, so they may look very different from each other.

LEDs are still a costlier option amongst different lighting source technologies and their long-lasting energy saving benefits must be carefully weighed against the investment costs.

#### 5.2.2.6 Lighting Ballasts and Drivers

The efficiency of the CFL lighting will depend on the ballast, which regulates the current going through the lamp and hence directly affect the energy use. The following types of ballasts can be used for energy savings (WBDG 2014):

- ***Rapid start ballasts*** are the most common type of fluorescent ballast. These ballasts offer a long lamp life at a reasonable cost. They have been used for years with lighting controls to provide energy savings.
- ***Instant start ballasts*** are usually the least expensive ballasts on the market. The efficiency of instant start ballasts is higher than rapid start ballasts, but lamp life is shorter, especially when the frequency of starts is increased due to the use of controls. They are often used where energy savings is the primary goal and lights are on continuously for very long periods of time. One advantage of the instant start ballast is that the lamps are wired in parallel, so that when one lamp on a multi-lamp ballast burns out, the others remain illuminated.
- ***Program start ballasts*** are some of the best to use for energy efficiency and long lamp life. These ballasts are slightly more expensive than standard rapid start ballasts, but use a "gentler" starting method so that frequent starting lessens the reduction in rated lamp life. These ballasts are recommended for smaller diameter fluorescent lamps and compact fluorescent lamps. With the right lighting controls scheme, program start ballasts can provide significant energy savings.

- ***Dimming electronic ballasts*** for linear fluorescent lamps usually fall into two categories. The first type has a dimming range of 5 % or 10 % up to 100 % light output and is generally the least expensive. This ballast is commonly used when the lowest light levels are not needed, or to achieve energy savings by dimming the lights when there is plentiful daylight. The second type of ballast, often referred to as an "architectural dimming ballast," is more expensive and has a dimming range of 1–100 % light output. This ballast is used in situations where lower light levels are desired.

***LED Drivers***

An LED driver regulates the power supply for an LED system, much like a ballast is to a fluorescent or HID lighting system. LED drivers may be constant voltage types (usually 10, 12 and 24 V) or constant current types (350, 700 mA and 1 A). Some drivers are manufactured to operate specific LED devices or arrays, while others can operate most commonly available LEDs. Drivers can enable dimming and color-changing or sequencing of LEDs. LEDs are easily integrated with circuits to control dimming and color-changing so that these functions can respond to preset commands or occupant presence or commands. Most LED drivers are compatible with commercially available 0–10 V control devices and systems such as occupancy sensors, photocells, wallbox dimmers, remote controls, architectural and theatrical controls, and building and lighting automation systems. Dimming does not result in a loss of efficiency of LED and in fact would result in better life expectancy due to lower operating temperatures.

### 5.2.3 Task Lighting

It is important to make a distinction between task and ambient lighting. Task lighting is the light actually used to perform a task, whereas ambient lighting is the general background illumination in the room. Task lighting should be free of distracting glare and shadows and should be bright enough to prevent eye strain. There is yet another type of lighting used by interior designers known as accent lighting or spotlight that is used to draw attention to certain objects in the room or outdoors such as painting, sculptures, landscaping and other priced possessions.

Often times, designers and engineers fail to make the simple distinction between general lighting and task lighting. This results in over-provision of lighting in many buildings. Addressing ambient and task lighting separately in a design can lead to significant energy savings. For example, effective daylight strategies can be used to provide general ambient lighting while task lighting could use more precise lighting technologies and controls.

### 5.2.4 *Lighting Controls*

Turning off lights when they are not being used and using as much daylight as possible reduces energy use. But achieving this solely through human behaviour change can be very tricky as its subject to complicated human factors and preferences. Hence technology such as timers and sensor-based controls is sometimes used to mimic the energy saving behaviour.

#### 5.2.4.1 Occupancy Sensors

Occupancy sensors help ensure that lights are only on when they are being actively used. Infrared sensors can detect heat and motion, and ultrasonic sensors can detect sound. Both must be installed correctly to ensure that they are sensitive to human activity rather than other activity in the vicinity (such as ambient noise).

#### 5.2.4.2 Photosensors

Photosensors use ambient light to determine the level of light output for a fixture. Photosensors might be used to turn outdoor lights off during daylight hours. The most common types of photosensors are the photodiode, the bipolar phototransistor, and the photoFET (photosensitive field-effect transistor). These devices are essentially the same as the ordinary diode, bipolar transistor, and field-effect transistor, except that the packages have transparent windows that allow radiant energy to reach the junctions between the semiconductor materials inside. Bipolar and field-effect phototransistors provide amplification in addition to their sensing capabilities.

Some studies have shown that utilizing occupancy and photo sensors can save an average of 30–40 % on energy costs (LBNL 2011). Sensors are not always able to detect and match the needs of the occupant. This is because sensors react to different wavelengths, such as visible light, ultraviolent radiation, and infrared radiation, and because they are often located far from the area of occupancy. Implementation of sensors in existing structures can be problematic because of the need for new fixtures, other wiring problems, and initial costs. Occupants may also object to automatic switch-off technology if it is poorly installed and is prone to premature switching; this can be remedied by more careful installation.

## 5.3 Energy Efficient Elevators

With the rise of skyscrapers and tall multi-storey buildings, vertical movement has become one of the key building provisions. Elevators can consume around 5 % of the total energy consumption in office buildings. Hence it is important to study the energy saving possibilities in elevators for green and smart buildings. A simple and passive measure to reduce elevator energy use is to encourage building occupants to use stairs. This however is not always practical especially for tall buildings and hence the focus on energy efficient elevators technology.

### 5.3.1 Gearless Traction Elevators

In a traction machine, hoisting ropes (or wire ropes), are attached to the top of the elevator and wrapped around the drive sheave in special grooves. The other ends of the cables are attached to a counterweight that moves up and down in the hoistway on its own guiderails. The combined weight of the elevator car and the counterweight presses the cables into the grooves on the drive sheave, providing the necessary traction as the sheave turns. To reduce the load on the motor, the counterweight is calculated to match the weight of the car and a half-load of passengers. As the car rises, the counterweight descends, balancing the load. This reduces energy consumption because the motor is required to lift no more than the weight of half a car load at any time.

Traditionally, electric traction lifts were equipped with DC motors due to their easy controllability, but the development of variable frequency drives led to the introduction of the now prevalent AC induction motors or permanent magnet DC motors. These drives provide excellent ride conditions, with smooth acceleration and deceleration and high leveling accuracy (Intelligent Energy 2010).

There are two main types of traction lifts: geared and gearless. Geared lifts use a reduction gear to reduce the speed of the car while in gearless lifts the sheave is directly coupled to the motor. The machine in gearless lifts consists of a motor, traction sheave and brake. Since the motor is directly coupled to the traction sheave, there are no transmission losses and they both rotate at the same speed.

### 5.3.2 Flat Steel Belts

Some traction elevators use flat steel belts instead of conventional steel ropes. These belts consist of a band of ultra-thin steel cables encapsulated in a polyurethane sheath. Flat steel belts are extremely light due to its carbon fiber core and a high-friction coating, and does not require any oil or lubricant. They are approximately 20 % stronger and can reach twice the life-time of the standard steel ropes (Intelligent Energy 2010). Flat belts allow the use of a much smaller diameter sheave (e.g. 100 mm) instead of the commonly used 720 mm diameter sheave. This way, at a given lift speed, the smaller sheave rotates several times as quickly as a larger sheave, so a smaller motor can deliver more torque to the load. Furthermore, it avoids the use of a gearbox which results in supplementary energy savings.

### 5.3.3 Machine-Room-Less (MRL) Elevators

The more energy efficient types of elevators are machine-room-less (MRL) traction elevators. Conventionally, all lifts, either traction or hydraulic, required a machine room where the motor and pump, in the case of hydraulic lifts, and a control cabinet were stored due to the size of the equipment. This machine room was typically located above the lift shaft for traction lifts (or below for hydraulic lifts). Manufacturers redesigned the motors (for example the use of compact permanent magnet motors as discussed in Sect. 5.3.6) and all of the other equipment normally housed in a machine room above conventional elevators to fit into the hoistway. These space-saving improvements eliminate the need to build and supply energy to a machine room and consume significantly less energy than the larger versions previously used and also generate less heat.

### 5.3.4 Regenerative Drives

In usual operation, when power flows into the motor, it creates a lifting torque on the shaft and elevator sheave, lifting the carriage. When the carriage travels down, the motor acts as a generator, transforming mechanical power into electrical power and pumping current back into the facility's electrical grid to use elsewhere. When a cab goes up with a light load and down with a heavy load, the system generates more power than it uses. Figure 5.15 shows an illustration of this concept.

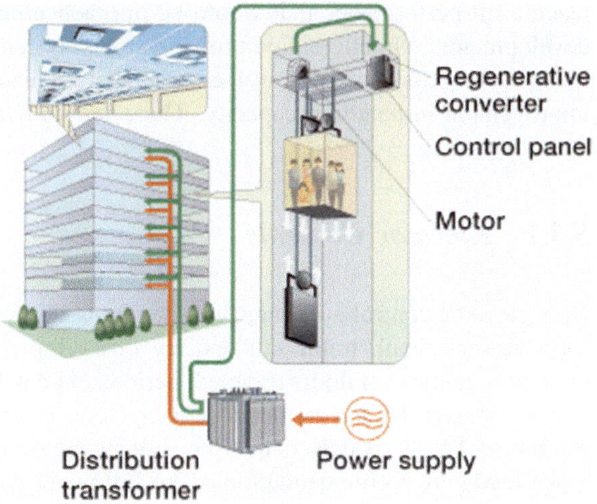


**Fig. 5.15** Regenerative drive elevators (Mitsubishi Electric)

Over time these small amounts of power generated during each elevator's sporadic decelerations add up to noticeable savings. They use less energy than non-regenerative drives, and reducing the excess heat in the building.

### *5.3.5 Permanent Magnet Motors*

Permanent Magnet Synchronous Motors (PMSM) are rapidly becoming the leading motor technology within the lift market. These motors do not have windings in the rotor and instead the magnetic field is provided by the magnets. Hence they have less Joule losses than induction motors and the magnetic losses in the rotor are also much reduced. Because it no longer requires an exciting current, PMSM have higher efficiency and present faster response speeds when compared to conventional induction motors. It presents many advantages, such as a simplified mechanical system for the lift, improved comfort, reduced noise and vibration, and energy savings (Intelligent Energy 2010).

### *5.3.6 Double-Deck Elevators and Twin Elevators*

One energy-saving change manufacturers have recently begun to offer is double-deck elevators. They are two cabs tall, one stopping at even-numbered floors and one serving odd. They can reduce a building's overall energy usage by reducing the number of stops and even the total number of elevators required when used with destination dispatch controls. When combined with hall call control systems, double-deck lifts performance is further improved, allowing buildings to reach a lift performance that would be impracticable with single-deck lifts. Recent developments in intelligent control and safety systems have allowed the use of two independent cars travelling in the same shaft. This can translate into a significant improvement in handling capacity. These are known as twin elevator systems.

### *5.3.7 Elevator Controls*

New elevator control software provides tools that elevator consultants use to perform elevator bank traffic studies. By observing the sporadic nature of elevator operation, number of floors traveled, periods of peak load, and the fact that elevators are not always loaded to rated capacity, consultants create energy consumption estimates. These models help a consultant create efficient control strategies and make hardware recommendations. The following controls can significantly reduce energy use in elevators:

- Software- and microprocessor-based controls instead of electromechanical relays.
- In-cab sensors and software that automatically enter an idle or sleep mode, turning off lights, ventilation, music, and video screens when unoccupied.
- Destination dispatch control software that batches elevator stops requests, making fewer stops and minimizing wait time, reducing the number of elevators required.
- Personalized elevator calls used with destination dispatch controls that eliminate the need for in-cab controls.

## 5.4 Plug and Process Load Management

Plug and process loads in buildings have a significant share of the building overall energy use. These are energy loads that are not related to general lighting, heating, ventilation, cooling, and water heating, and that typically do not provide comfort to the occupants. Rather these loads are generated from ‘plugged-in’ equipment (into wall outlet plug) and used by the occupants such as refrigerators, microwaves, computers, printers, projectors, audio equipment, coffee machines, televisions, chargers, etc.

There are huge opportunities in understanding and managing these loads. Not only do they save energy directly, but cooling energy is also saved by not removing the heat generated by these plug loads. The main challenge with managing plug loads is that their use is governed by occupants with varying needs and unknown use patterns.

### *5.4.1 Use of Energy Efficient Equipment*

There are now several equipment that consume much less power for the same functionality. The conscious choice of such equipment can help in reducing plug power from day one. In many countries such equipment are rated and labeled based on their energy efficiency. The users can simply choose the equipment by looking at the energy efficiency label and making a conscious decision considering other factors such as price, functionality and suitability.

For example, the European ENERGY STAR program is a voluntary energy labeling scheme for office equipment. With the ENERGY STAR logo, consumers can easily identify energy efficient products. Similar labels are also practiced in many other countries where stars, ticks or color coded bars are used to easily identify the energy efficiency of purchased equipment in stores. Figure 5.16 illustrates an example of the EU energy efficiency logo used to rate equipment.

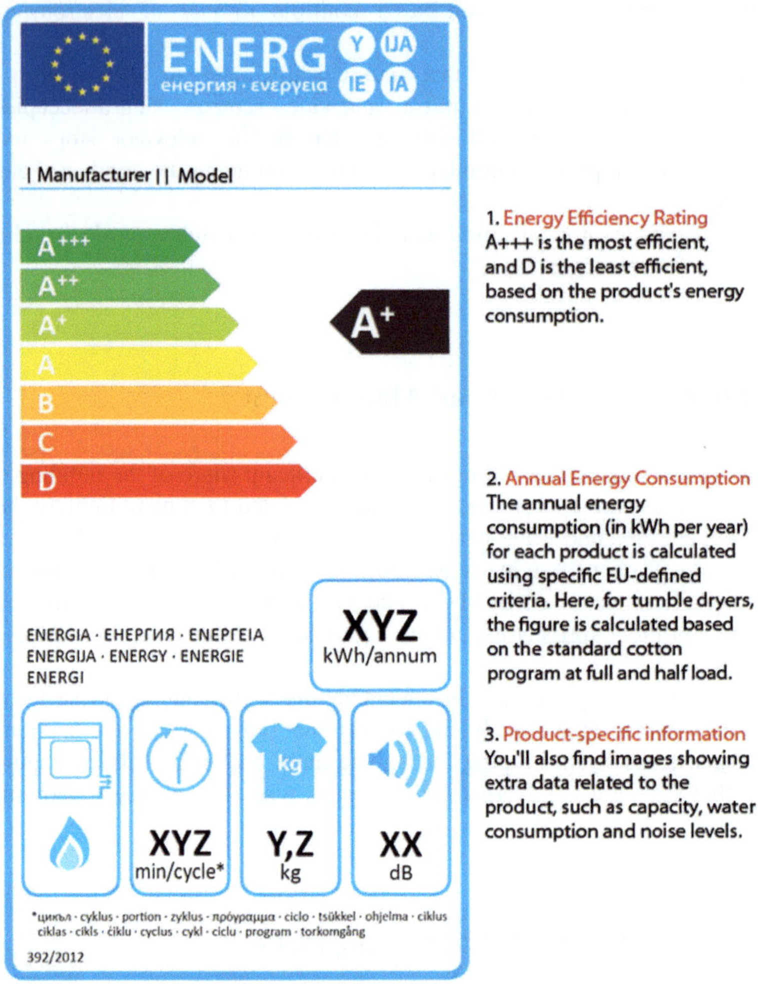


**Fig. 5.16** EU energy efficiency label example (*Source* which.co.uk)

## 5.4.2 Timer Controlled Plugs

Several control strategies can be employed to turn off devices when not in use and significantly reduce energy consumption. Timers and timer plug strips are unobtrusive to the participants and reduced electricity use significantly, making them good options to control devices with regular schedules. There are now simple manual timer control plugs available that can be 'programmed' for daily operation between certain hours of the day. There are also digital time switches available that can be programmed more accurately to consider weekends and hourly variations during the month. Figure 5.17 shows photos of such timer controlled electric plugs.

Fig. 5.17 Manual (*left*) and digital (*right*) timer controlled electric plugs

### 5.4.3 *Smart/Advanced Plug Strips*

Smart power strips employ smart control strategies to shut off the supply power when electronics are not in use or not required to be used. Plugged-in devices consume some power even when they are on the stand-by mode. According to Lawrence Berkeley National Laboratory, this standby power draw, or phantom load, can account for 5–10 % of a typical home's annual electricity use. Earle and Sparn (2015) classify these advanced power strips into different types as described below. A photo of various smart power strips available in the market is shown in Fig. 5.18.

#### 5.4.3.1 Current-sensing Power Strips

A current-sensing power strip monitors the power draw of the controlled outlets to determine whether the connected devices are on or off. If they are off, the power smart strip shuts off the supply power to the controlled outlets. Some action must then be taken by the user (e.g. pressing a button) to restore power to those outlets. The algorithm that the smart strip uses to determine when the controlled appliances are "off" varies. Some algorithms look at peak power use for all the connected devices and shut off the supply to all outlets when the total power has dropped by a certain percentage. Others monitor smaller groups of outlets and control each group independently. Such strips can essentially eliminate the electricity consumption of appliances in stand-by mode.

Fig. 5.18 Different types of smart/advanced power strips (*Source* Earle and Sparn 2015)

#### 5.4.3.2 Master-controlled Power Strips

A master-controlled (also called master-slave) power strip uses the power state of the "master" appliance to determine whether the supply power to the controlled outlets (where peripheral devices are plugged in) should be shut off. It works on the (usually accurate) assumption that, for example, you do not need your DVD player (peripheral) to be on if your TV (master) is off. The master appliance (your choice, but typically the TV or computer) is plugged into a special outlet that does not turn off. The controlled outlets are powered only when the master appliance is in use. Most master-controlled power strips use current sensing to determine the power state of the master appliance, and many have adjustable current thresholds so that you can control the load level that should be considered "off".

#### 5.4.3.3 Infrared (IR)-based Activity Monitor Power Strips

A simple IR-based power strip uses the IR signal to activate a switch that turns the controlled outlets on and off; it is basically a traditional power strip with a remote switch. IR sensing can also be coupled with a current-sensing power strip. In this case, the IR signal "wakes up" the power strip and restores power to the controlled outlets.

#### 5.4.3.4 Motion-sensing Power Strips and Activity Monitors

A motion-sensing power strip turns off the controlled outlets when it detects no motion in the room for some period of time. A motion-sensing power strip turns off power to appliances when they are on but are no longer in use by assuming that motion is a good indicator of device usage. An activity monitor may employ a more complex suite of sensors than a motion detector to decide on cutting-off power to devices.

#### 5.4.3.5 Remote-switch Power Strips

The on-off switch on a traditional power strip may at times be hard to reach e.g. under the table or behind the computer workstation. A remote switch uses IR or radio signals to allow users to turn outlets on and off remotely. Some products have a subset of remotely controllable outlets on an otherwise conventional power strip.

#### 5.4.3.6 Timer-controlled Power Strips

A power strip with a timer switch can be programmed to shut off supply power during times when the devices are not to be used. These are similar to the timer controlled plugs discussed in the earlier section. For enhanced automation, a timer switch can be integrated into a current-sensing power strip, where the timer starts counting down once its determined that the appliance is in the stand-by mode.

### *5.4.4 Software Power Management Settings*

Enabling and properly programming existing power management settings of computers and imaging equipment provides the largest energy savings opportunity. If adequate software is already installed on the system, this solution can be implemented at no cost. There are barriers to be addressed if energy savings are to be achieved, such as a lack of user information and education, users requesting remote access to their desktop computers and conflicting practices with existing IT management policies. Alternatively, low-cost, network-based power management software that allows IT managers to centrally control power to devices during nights and weekends may be purchased.

## References

Conant A et al (2004) Comprehensive diagnostic and improvement tools for HVAC-system installations in light commercial buildings. Proctor Engineering Group. Final Report. 31 Oct 2004. NETL. DE-FC26-01NT41256

Dieckmann J, Roth K, Brodrick J (2007) Emerging technologies—magnetic refrigeration. ASHRAE J 74–76

Earle L, Sparn B (2015) Managing plug loads with advanced power strips. Home Energy 32(4)

EcoCampus (2016) Liquid immersion cooling system. Available online at: http://ecocampus.ntu.edu.sg/Pages/Liquid-Immersion-Cooling-System.aspx. Accessed: 15 Mar 2016

EERE (2011) Energy savings potential and RD&D opportunities for commercial building HVAC systems. In: U.S. Department of Energy, Energy Efficiency and Renewable Energy, Building Technologies Program (ed)

EERE (2012) Life-cycle assessment of energy and environmental impacts of LED lighting products: Part I: Review of the life-cycle energy consumption of incandescent, compact fluorescent, and LED lamps. In: Energy, U. S. D. O. (ed)

Energy Star (2016) U.S. EPA/DOE Energy Star Program. Learn about compact fluorescent light bulbs. http://www.energystar.gov/index.cfm?c=cfls.pr_cfls_about. Accessed 3 Jan 2016

Goetzler W et al (2009) Energy savings potential and R&D opportunities for commercial refrigeration. Final Report. Navigant Consulting, Inc. 23 Sep 2009

Hamilton S et al (2003) Improved Duct Sealing. Emerging Technologies. ASHRAE J

http://www.gelighting.com/LightingWeb/na/solutions/technologies/

Intelligent Energy (2010) Europe, energy efficient elevators and escalators. Available online at: https://ec.europa.eu/energy/intelligent/projects/sites/iee-projects/files/projects/documents/e4_publishable_report_en.pdf. Retrieved 10 Sep 2015

Intergovernmental Panel on Climate Change (IPCC) (2001) Global warming potential of refrigerants. Retrieved: http://www.ipcc.ch/ipccreports/tar/wg3/index.php?idp=144

Kristoferb (2016) At English Wikipedia [CC BY-SA 3.0 (http://creativecommons.org/licenses/by-sa/3.0) or GFDL (http://www.gnu.org/copyleft/fdl.html)], via Wikimedia Commons. Accessed 15 Feb 2016

Landis GA (2016) Photograph uploaded on Wikipedia and distributed under the Creative Commons License. Available at: https://commons.wikimedia.org/wiki/File:LED_bulbs_2012.jpg. Accessed 15 Feb 2016

LBNL (2011) A meta-analysis of energy savings from lighting controls in commercial buildings. Performed by Lawrence Berkeley National Laboratory and Erik Page & Associates, Inc.

LBNL. Aeroseal duct sealing. LBNL—from the lab to the marketplace. http://eetd.lbl.gov/l2m2/aerosol.html

Mitsubishi Electric, Elevators—using energy wisely. Available online from: http://www.mitsubishielectric.com/whatschanging/ecochanges/elevators/. Accessed 15 Sep 2015

Nevit (Own work) [CC BY-SA 3.0 (http://creativecommons.org/licenses/by-sa/3.0)], via Wikimedia Commons. Accessed 25 Mar 2016

SGL Carbon (2012) http://www.sglgroup.com/cms/international/products/product-groups/eg/specialty-graphites-for-construction-hvac/

United Nations Environment Programme (UNEP) (2007) Information about the management and phase out of hydrochlorofluorocarbons (HCFCs). Retrieved: http://www.uneptie.org/ozonaction/topics/hcfc.asp

WBDG (2014) (Whole Building Design Guide) A program of the National Institute of Building Sciences. Available online at: https://www.wbdg.org/resources/efficientlighting.php (updated 25 august 2014). Accessed 14 Nov 2015

# Chapter 6
# Building Management and Automation Technologies

Building Management Systems (BMS) or Building Automation Systems (BAS) are computer-based control system installed in buildings that monitor and control the building's mechanical and electrical equipment such as ventilation, lighting, power systems, fire systems, and security systems. They collect and analyze data and then provide insights or take necessary actions to improve the building's efficiency and productivity. A BMS essentially consists of two components: hardware and software. The hardware used is that for data collection (analog and digital meters and sensors), data analysis (computers or servers and dashboards for visualization), controllers or actuators. The software program serves as an interface between measurements taken and the building manager or controllers. A layer of software applications is also often used in combination and these applications defines the control strategies (HVAC controls, lighting controls, etc.), instruction and optimization routines.

Brambley et al. (2005) describe the conceptual framework of a Building Automation System as having four areas: applications, hardware, communications, and oversight. This framework is summarized in Fig. 6.1.

## 6.1 Sensors and Meters

Sensors are currently prevalent in several building control applications. One of the most important types of sensor for monitoring energy efficiency and for fault detection is an electrical power/current meter. The most commonly used environmental sensors are those for measuring temperature, relative humidity (RH), carbon monoxide (CO), and carbon dioxide ($CO_2$). To measure temperature, both mechanical (e.g., thermally expanding metallic coils) and electrical means (e.g., thermistors, metallic RTDs, thermocouples, digital P-n junctions, infrared thermocouples) provide sufficient accuracy for current needs and have proven to be fairly reliable. The most common RH sensors yield an output that is proportional to

 
N.Y. Jadhav, *Green and Smart Buildings*, Green Energy and Technology,
DOI 10.1007/978-981-10-1002-6_6

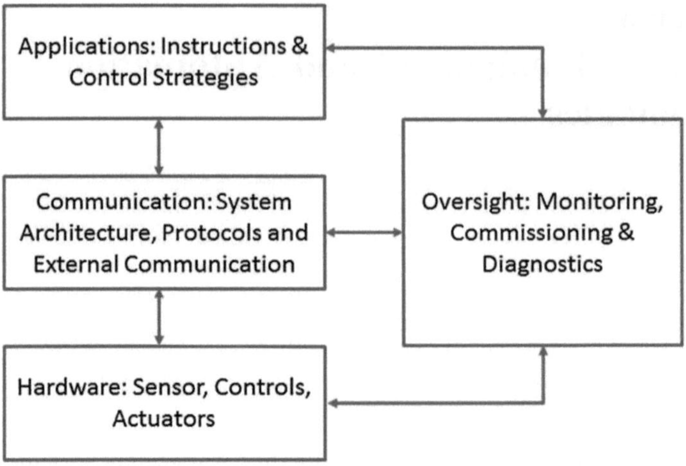


**Fig. 6.1** Conceptual framework of a BMS/BAS (Brambley 2005)

either the capacitance or resistance of a hygroscopic material. The commonly $CO_2$ sensors contain a small cell in which infrared light is passed through the air sample and the absorption of the infrared light can be related to the $CO_2$ concentration.

Occupancy sensors are valuable tools for energy conservation that are used extensively to turn off lighting and other equipment when no occupants are present. Infrared detection of people or ultrasonic detection of movement have been the predominant means of achieving such detection. Additional lighting control can be achieved through photosensors by monitoring the amount of daylight so that lighting levels can be suitably adjusted.

### *6.1.1 Smart Sensors and Networks*

Smart Sensors have on board micro processing for data analysis before data transmission and can also participate in sensor networks combining raw data from multiple sensors to develop higher level information at sensor node. This hence makes the sensor node more powerful in terms of decision making ability and initial analysis. The data from a sensor network can be analyzed to understand consumption patterns, occupancy schedules and equipment operational efficiency in order to fine tune the overall effectiveness of the building controls. There are also sensor nodes that can perform sensing of various parameters via integration of different sensor technologies (e.g. temperature, humidity, light, occupancy, etc.) into one sensor node. Such sensors nodes are referred to as 'multi-modal smart sensor nodes'.

### 6.1.2 Wireless Sensors Networks

The key promise and opportunity of wireless technology in building operation is to significantly reduce the cost of installing control systems by eliminating the control wires. The availability of low-cost wireless sensor and control systems could not only reduce installation costs overall, but also lead to increased use of sensors and control devices necessary to establish and maintain highly energy-efficient building operations and productive and healthy work spaces.

Early adopters of wireless technologies are Johnson Controls, Siemens, and Delta Controls offering wireless temperature sensors that are either fully integrated into wired building control networks or provide non-networked direct communication between devices. Wireless lighting switches are offered by Lutron Electronics Co. (Lutron 2003) and others. Wireless and multi-modal sensor nodes that can perform several sensing functions are also now available.

Typically, wireless sensors need to use batteries to provide the small amount of power required for the sensing function as well as most importantly the data communication. This can be one of the major problems in wireless sensors as although they reduce or eliminate the wires, they are not maintenance free as the battery life determines their useful lifetime before being maintained. Changing batteries of several wireless sensors in a building can be a major hassle. Hence there are now self-powering or energy harvesting wireless sensors available that can integrate some energy generation component such as a solar cell, thermoelectric or piezoelectric generators. Figure 6.2 shows a smart, wireless sensor node with solar energy harvesting capability.

### 6.1.3 Integrated Communication Technologies

Information provided by smart sensors and smart meters needs to be transmitted via a communication backbone. This backbone is characterized by a high-speed and

**Fig. 6.2** Smart wireless sensor with solar energy harvesting capability (EnOcean)

two-way flow of information. In case of wireless sensor networks the use of very low power methods for radio communication and data acquisition is highly important in order to conserve the battery. In many applications, a wireless sensor network communicates with a Local Area Network (LAN) or Wide Area Network (WAN) through a gateway. The Gateway acts as a bridge between the WSN and the other network. This enables data to be stored and processed by devices with more resources, for example, in a remotely located server.

A Local Area Network (LAN) can be formed by communication technologies classified broadly into three main categories:

(1) wireless
(2) wired Ethernet
(3) in-building Power Line Communication (PLC).

Wireless technologies are popularly known by the classifications done by the IEEE (Institute of Electrical and Electronics Engineers) standards. The popular wireless IEEE standards are as follows:

- Wi-Fi (IEEE 802.11),
- WiMAX5 (IEEE 802.16),
- ZigBee (IEEE 802.15.4) and
- Bluetooth (IEEE 802.15.1).

Wide Area Network (WAN) technologies are important is the building information needs to be transmitted over a long distance typically for central management at an building estate level or interaction with the smart grid. There are several WAN technologies available such as follows:

- Asymmetric Digital Subscriber Line (ADSL)
- Cable Modem
- Fiber broadband or Fiber to the Home (FTTH)
- Power-line communication (PLC)
- Cellular services
- Satellite services
- Paging systems.

The choice of WAN technologies will depend on factors such as reliability, low-cost, security and the network infrastructure that is already available. Typically, WAN deployment is done by a service provider such as a tele-communication company. There are also now low power WANs available such as those made available by Sigfox and the LoRa Alliance that are a type of wireless telecommunication network designed to allow long range communications at a low bit rate among connected devices such as sensors.

## 6.2 Controllers

Controllers are the devices and algorithms that take data from sensing and data acquisition systems and convert it to desired actions and signals to instigate action by actuators for controllable devices, such as dampers, valves, on/off switches, and other devices whose states or modes of operation can be varied. Most controllers in a typical BMS take three general forms:

(1) scheduling and mode selection,
(2) pneumatic controllers and
(3) direct digital controllers (DDC): proportional, integral, derivative (PID) control, and

### *6.2.1 Mode Selection Controls*

Mode selection is usually done on a schedule based on time of day or day of year. For example, air handling fans in an office building may be scheduled for certain hours when occupants are expected to be present, say 7 a.m. to 7 p.m. on weekdays, and off at all other times. In some cases, more advanced control selection of modes may be used, for example, air-handling fans might be turned on based on data from occupancy sensors.

### *6.2.2 Pneumatic Controls*

Pneumatic controllers were used for many years in commercial buildings to control modulating devices. Over the last two decades, direct digital control has gained favor for new installations, but many buildings still control devices, especially terminal units, using pneumatic control.

### *6.2.3 DDC Control Mode*

Direct digital control is commonly used in modern building systems to control devices whose load changes over shorter periods of time than a day. These devices can be on/off controlled, like thermostats in homes, or modulated, such as a chilled-water valve in some air-handling units. PID control is linear by definition, even though many devices to which it is applied in buildings are non-linear. In most cases, during normal operation, this does not present a problem because the non-linear devices and loops can be considered linear over ranges of operation. The PID control action is used to speed up the response of the system to reach the set point value in case of sudden changes.

### 6.2.4 Advanced Intelligent Controls

There are now advanced control systems available that have humanlike capabilities such as reasoning, pattern recognition, adaptation and learning. Rule-based expert systems are based on human like reasoning and can be powerful in solving problems related to self-tuning, fault diagnosis, scheduling and planning. Another advanced control method is the use of fuzzy logic in controlling complex systems with imprecise or uncertain system information and behaviour. The use of artificial neural networks in controls is also becoming popular as well trained neural network models can have powerful functions in learning and self-organization based on both qualitative and quantitative knowledge.

### 6.2.5 Model Predictive Controls

Model predictive control (MPC) is an advanced method of process control that is based on constrained optimal control. In building control for example, it could be used for optimizing the energy delivered (or cost of the energy) subject to comfort constraints. MPC relies on having a dynamic model of the building and simulating performance of the building based on this model and input parameters such as weather and occupancy. Typically, the MPC routine is run over a period of time or control interval e.g. every 30 min and provides the optimal inputs or set-points for the DDC controller to take action for the next 30 min. The measured performance based on control action is then fed again to the MPC controller at the start of the next time interval along with the time varying parameters such as energy prices, occupancy and weather predictions. The basic principles and building block of the MPC system are shown in Fig. 6.3.

## 6.3 Actuators

Actuators are the portion of a control system that can control or vary the output of a physical device or process, such as a relay, variable damper, motor, or lighting ballast. Actuators are usually closely coupled or directly attached to the equipment

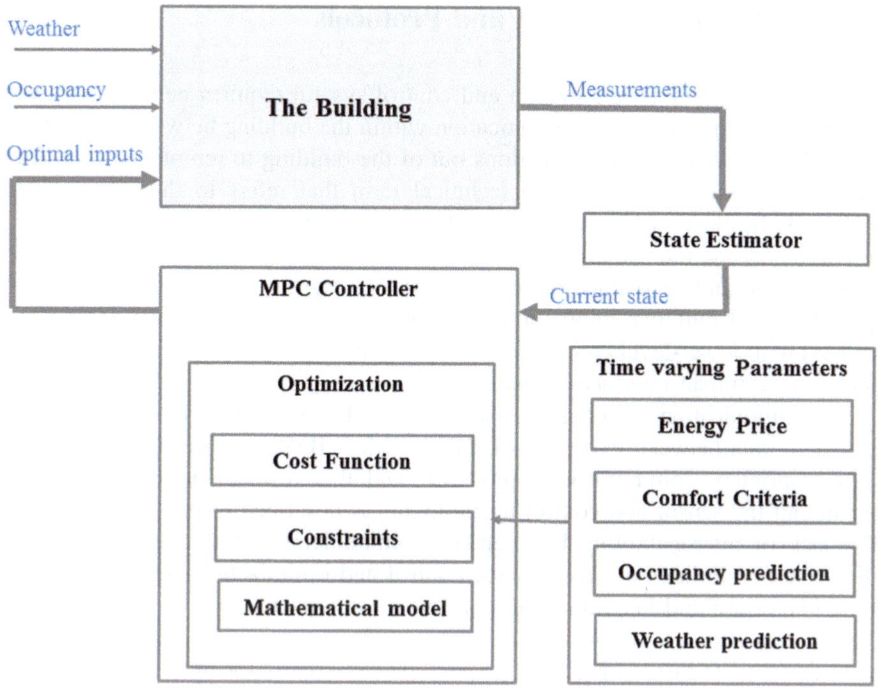


**Fig. 6.3** Basic principles of MPC for buildings (Cigler 2013)

they control, thus the electrical integration of actuator to physical equipment is usually straightforward. Actuators usually do not embed intelligence; intelligence usually resides in the controller, which is often separated from the actuators by a considerable distance.

In the analog mode of operation, actuators are generally directly wired to the physical system controlled. The control signal that dictates the specific state of the actuator (open, closed, etc.) is usually transmitted from a physically remote controller using low-voltage wiring. In conventional control systems, many actuators are operated by one controller output. In direct digital control (DDC) systems, which are increasingly common for HVAC applications and are beginning to be considered for lighting systems (for example DALI—Digital Addressable Lighting Interface), actuators are digitally operated, often from remote field panels that contain a controller as well as digital connections to other field panels. Digital actuators can be operated from simple field busses (multiple actuators connected to a single pair of wires and controlled digitally), which reduces the wiring problems of analog actuators.

## 6.4 BMS Communication and Protocols

The advanced building automation and control system requires network hardware and protocols that support communication within the building between devices and controllers as well as communications out of the building to remote sites, utilities, and other service providers. The technical term that refers to this data transfer system is the 'bus', which comprises of the data transfer hardware (cables, etc.), software and communication protocol. Most building automation networks consist of a primary and secondary bus which connect high-level controllers (generally specialized for building automation, but may be generic programmable logic controllers) with lower-level controllers, input/output devices and a user interface (also known as a human interface device).

A communication protocol consists of a set of rules that have to be applied to exchange data between two parts of a BMS. Most BMS manufacturers have their own proprietary communication protocols and that restricts building owners in expanding the sensor and control technologies as described in the earlier sections. The lack of interoperability between the communication protocols can severally affect the possibilities of facility wide control and optimization especially considering expansion and integration of smart wireless sensor networks.

In recent years, a number of open communication protocols have been developed that enable interoperability by specifying standards for how the bus devices interoperate e.g. by defining the of services that are used for device and object discovery and data sharing. The following the popular open protocols for BMS/BAS used today and there further are more under development that are not listed here:

- BACnet (Building Automation and Control networks) is an ASHRAE, ANSI, and ISO 16484-5 standard protocol.
- LonWorks (local operating network) is a networking platform built on a protocol created by Echelon Corporation for networking devices over media such as twisted pair, powerlines, fiber optics, and RF. It has been accepted and adopted by the international standards organizations (ANSI/CEA 709.1 and IEEE 1473-L).
- Modbus is one of the most widely used protocols in the industrial manufacturing environment and also a useful tool for achieving interoperability in building automation applications.
- KNX is a standardized (EN 50090, ISO/IEC 14543), OSI (Open Systems Interconnection)-based network communications protocol for building automation.
- S-Bus (Smart-BUS) is a protocol based on RS485 connection Topology for intelligent buildings.
- Open Platform Communications (OPC) is a series of standards and specifications for industrial telecommunication.

## 6.5 Monitoring and Diagnostics

The data from the various sensors and meters in the building can provide very useful insights into the building's operational performance and possibilities for optimization and energy savings. The data collected is useful to ensure that critical building services are well maintained and any faults in the system can be diagnosed in a timely manner.

### *6.5.1 Dashboard and Data Analytics*

Dashboards are visualization tools that provide useful information to the user at a glance. Dashboards are mostly used by facility managers to get a quick overview of building operations, faults and quick diagnostics on the performance. However, apart from building facility managers, there are other stakeholders in the building that can look at the building data for different purposes. These stakeholders include the building owners, executive management, occupants, tenants, visitors, regulatory officers and researchers. Dashboards can be useful visual tools to make the occupants aware of the building sustainability issues and influence their behaviour.

It important that the dashboard is customizable to individual needs and conveniently available. Online or web-based dashboard can be conveniently accessed by individual users and they should be compatible with various view devices such as computers, tablets and smart phones. Also the dashboard should be able to arrange data well and provide a drill-down of the data as and when required, while maintaining the high-level metrics and performance indicators. Figure 6.4 shows an example of a building monitoring dashboard.

**Fig. 6.4** Example of building monitoring dashboard (CCS 2013)

Analytics take data from a wide variety of sources and turn it into actionable information. Essentially analytics transform data into metrics and can provide useful insights on how and when buildings meet the pre-defined performance metrics. Analytics should allow historical tracking of performance and trend analysis to detect anomalies in performance. Analytics can help pinpoint which systems and equipment have irregularities or faults, with prioritization based on performance metrics such as energy, cost, severity, and comfort impact. Analytics should also help find hidden system inefficiencies with effective performance benchmarking.

The main challenge with data analytics in buildings is that the data from building is often not structured and can come from various systems that have their own accuracy and reliability issues. An example of unstructured data in buildings is it maintenance data and also retrofit data is in most cases not well structured. Such data sets are complex, massive and in a constant state of change and are referred to as 'Big data'. Big data analytics in buildings is an important research topic at this moment and with the increasing availability of commercial analytics engines and computational power, it will be a key strategy for smart buildings of future.

### *6.5.2 Alarm Management*

The Building management system usually generates numerous alarms to highlight faults and operational/maintenance issues. Alarms range from major issues, such as a power outage, to insignificant messages, such as a notification that a self-test has started. By prioritizing and structuring the numerous notifications generated by building systems, a smart building solution focuses engineers' attention on the most critical events. They can concentrate on urgent and impactful interventions from the perspective of occupant comfort, energy consumption, cost and business impact.

### *6.5.3 Automated Fault Detection and Diagnostic*

Automated fault detection and diagnostics (AFDD) is an emerging smart building solution and has tremendous potential to improve the operational performance of buildings. The AFDD software and tools should autonomously and in real–time, acquire and analyze data from control hardware and instrumentation products found in buildings. These tools help automatically detect faults in equipment and building system by using artificial intelligence, deductive modeling and statistical techniques. Upon detecting a suspected fault, AFDD tools will further provide expert and timely diagnosis of the underlying cause, helping maintenance staffs ensure that buildings perform effectively and efficiently.

## 6.6 Application and Control Strategies

The BAS or BMS system needs to have a layer above the hardware and software interfaces as described above, that deals with applications and control strategies. The applications could be based on user-interactivity, security, fault detection or energy and other services management. The control strategies would be based on observations and data analysis done in these applications. This layer of applications and control strategies determines the overall "smartness" of the building. Figure 6.5 shows and illustration of how smart building applications can enhance occupant comfort, convenience and overall sustainability performance of the building.

**Fig. 6.5** An illustration of how smart building applications can take automatic actions to optimize energy costs and occupant comfort (*Source* Honeywell)

## 6.7 Internet of Thing (IoT) Enabled Building Management Systems

Just as the internet enables computers to be connected, the Internet of Things (IoT) enables physical objects to be connected on a network that can exchange data between the connected objects. Objects such as devices, vehicles, buildings, etc. would have embedded electronics or sensors that would make them addressable or to be 'found' on the internet of things via wired or wireless internet connections.

IoT can enable the buildings to be managed with the help of low-cost sensors that collect data and from building spaces (indoor environmental conditions) and equipment (lighting, air-conditioning, etc.) and analyze the data using a data analytics engine sitting in the 'cloud' (i.e. connected via the internet). This hence can shift the control capability and intelligence from the traditional BAS/BMS to a cloud-based management platform that can then communicate with the actuators to execute the control actions. With the low cost of sensors and communications devices such as a gateway, this means that buildings can be automated at a much lower cost compared to the traditional and proprietary BAS/BMS.

Beyond the simple control capabilities, the internet platform can also help in optimizing buildings for enhanced energy operations. It can also gather real-time data from the occupants about their presence in the buildings and hence fine-tune the occupant comfort while conserving energy. For example, the cloud-based building management system can collect data from the occupant's schedule planning tool or digital calendar to know if the occupant will be at his/her desk at a particular time and based on this information can automatically switch-off lights and air-conditioning as appropriate. This can then be verified using the wireless temperature and light sensor installed in the room.

The following are the key components of an IoT enable building management system:

1. Directly cloud connected devices e.g. smart phones and smart meters.
2. Sensors that collect data from the objects and spaces.
3. Gateways that transmit the data to the cloud (could be a corporate cloud i.e. intranet or the open internet).
4. Networking and data storage:
    a. communication between the sensors and gateways and the gateway and the server.
    b. The data center or the server that collects and holds the data as well as provides the computational hardware.
5. Software and data analytics engine that makes sense of the data, compute optimal operation parameters and help the occupant and facility managers visualize the data appropriately.
6. Actuators that are connected to the internet via the gateway.

# References

Brambley MR et al (2005) Advanced Sensors and Controls for Building Applications: Market Assessment and Potential R&D Pathways, PNNL-15149 (U.S. Department of Energy), April 2005

CCS (Continental Control Systems LLC) (2013) http://www.ccontrolsys.com/w/Newsletter_-_Spring_2013. Accessed 28 Nov 2015

Cigler J (2013) Model predictive control for buildings. Doctoral thesis, Czech Technical University in Prague, June 2013

EnOcean, https://www.enocean.com/de/enocean_module/stm-310/

Honeywell, https://buildingcontrols.honeywell.com/buildingintel/2015/09/graphic-24-hours-in-a-smart-building/. Retrieved 12 Mar 2016

IEEE Standards, http://standards.ieee.org/

Lutron (2003) Product information available at the website: http://www.lutron.com/radiotouch/

OECD, Smart sensor networks: technologies and applications for green growth, Dec 2009. https://www.oecd.org/sti/ieconomy/44379113.pdf

# Chapter 7
# Renewable Energy Integration in Buildings

In the earlier sections we looked at energy efficiency in buildings and how technologies and tools can help reduce the energy use of buildings drastically. However, it's not possible to use absolutely zero energy in occupied and operational buildings. The term 'zero energy building' is possibility when there is enough on-site energy generation that meets the energy needs of the building. In fact, the definition of a zero energy building is still being debated around the world. The Global Building Performance Network (GBPN) has devised several definitions of Zero Energy Buildings based on a spectrum approach with increasing demands on how efficiently the building utilizes and generates energy as follows (GBPN 2016):

1. ***Nearby Zero Energy Building***: A nearby zero energy building is a building with extremely low energy consumption that is close to achieving net zero energy. Buildings should be shown to be as energy efficient as possible before renewable energy systems are added to the building.
2. ***Net Zero Energy Building***: A net zero energy building is a building that by and large produces as much energy from renewable sources as it consumes to achieve appropriate thermal comfort levels. Energy consumed must be produced on-site by renewable sources. Onsite refers to energy sources that are in, on, under, or adjacent to the building. Adjacent can include energy generated by a system within a reasonable distance from the building or closed grid. The building may form part of a larger development such as a small housing development.
3. ***Positive Energy Building***: A positive energy building is a building that produces more energy from renewable sources than it consumes (excluding plug loads) to achieve appropriate thermal comfort levels. Renewable sources include Solar thermal, photovoltaic, wind and geothermal. Hydro may also be included if locally and sustainably sourced (with security of supply).
4. ***All Energy Positive Building***: An "All Energy Positive Building" is a building that produces more energy from renewable sources than it consumes to achieve appropriate comfort levels. An all energy positive energy building can be



N.Y. Jadhav, *Green and Smart Buildings*, Green Energy and Technology,
DOI 10.1007/978-981-10-1002-6_7

distinguished from a "Positive Energy Building" by virtue of the fact that it includes all energy loads in the building.

5. ***Zero Energy Footprint Building***: A "Zero Energy Footprint Building" is an all energy positive building that produces more energy from renewable sources than it consumes to achieve appropriate comfort levels. The surplus must cover the energy used during the construction of the building.

It is however not good enough to have any kind of energy generation in buildings for it to be called a zero energy building. The energy generation in building has to be based on renewable energy sources that are available within the building and the building should not rely on getting the energy generation sources from outside the building. Renewable energy is a growing energy generation form worldwide and it is estimated that by the year 2040, about 60 % of electricity generated can be attributed to renewable energy (BNEF 2016).

There are various forms of renewable energy generation options such as solar, wind, geothermal, hydropower, biomass, etc. Integrating renewable energy in buildings is a challenge as the space and provisions for the same can compete with other functional requirements. Also, it's not possible to integrate all forms of renewable energy available today into buildings. In this chapter we focus on renewable energy technologies that can be well integrated into buildings as energy generation sources.

## 7.1 Solar PhotoVoltaic (PV) Energy

The sun's energy can be converted to make electricity using solar photovoltaics (PV). Solar PV cells convert sunlight directly into electricity by using the photovoltaic effect. Sunlight is abundantly available in many parts of the world and in fact some building design strategies we discussed in Chap. 4 are focused on reducing direct sunlight and heat gains in buildings. Hence solar PV is a good fit for renewable energy generation in buildings.

There are now various solar PV cell technologies commercially available and the most prominent one is the silicon-based solar cell technology. Silicon-based solar cell technology is often referred to as first generation solar PV technology and has seen a rapid adoption worldwide for power generation as its cost has fallen drastically over the past two decades. Second generation solar cells are based on what is called a 'thin-film' solar cell technology, which is relatively easier to manufacture can be flexible and light-weight. There are also third generation solar cells now investigated that rely on novel nano-structured materials that have a promise of very high efficiencies, but are still in the research and development stage of development.

The performance of a PV system is influenced by the amount of solar energy available at a specific location and by the effectiveness of the system to convert solar energy to electrical energy. The focus of this section is on application of solar PV technology in buildings and hence we look at two possible applications of solar PV in buildings, viz. rooftop PV and other building integrated PV (BIPV).

### *7.1.1 Rooftop Solar PV*

Solar PV cells are formed into modules or solar panels that can be conveniently used to harness energy from the rooftop. Arrays of solar panels can be connected together electrically to generate clean energy from the available rooftop area. The solar panels need be oriented properly to get enough sunlight by considering the sun path e.g. towards south in the northern hemisphere and towards north in the southern hemisphere. They are also tilted or inclined at an angle equal to the latitude of the location.

Solar panels are secured to the roof using mounting systems that can be attached to the roof. There are different ways of mounting the solar panels on rooftops for different types of roof e.g. for metal roofs, the mounting structure is bolted in or clipped-on to the roof and in case of flat concrete roofs, a ballast system can be used to mount the modules on top without the need for drilling. While PV systems add a relatively low additional load on a roof, it is still important to ensure that the overall system is in line with structural allowances, and that it does not compromise the building's weather-proofing.

Solar panels installed on the rooftop, silently convert sunlight into electrical energy and have no moving parts. This means that there is very little maintenance and they can last very long. Commercial solar panels have already been tested for more than 20 years of lifetime. There is however a small degradation in performance that happens over the years (at a rate of approximately 0.3–1 % per year, depending on the module type and local conditions) and designers need to factor this in appropriately. Inter-row shading and the surrounding landscape influences how much exposure the system has to the sun. Neighbouring buildings, trees or natural features can shade part or the whole of a system, affecting overall energy generation. Figure 7.1 shows an aerial picture of buildings with rooftop Solar PV installation at the NTU Singapore campus, which boasts one of the largest rooftop solar PV installation of about 5000 kilo-watts-peak (kWp) power.

Apart from solar panels, there are other essential components in the rooftop PV system that are often termed as Balance of System (BOS) components that include the following:

- **Racking and mounting systems** that help secure the solar panels to the rooftop.
- **Inverters**: they convert the direct current (DC) electricity generated from the solar panels into alternating current (AC) electricity that can be directly used in the building. A typical design has central inverters or string inverters to aggregate and convert the output of individual PV modules. Micro-inverters can also be integrated within a module. This to convert DC power to AC power at

**Fig. 7.1** Photo of solar panels installed on the rooftop of the NTU campus buildings in Singapore

the module itself. Each module can therefore be monitored and controlled individually. This could result in the entire system having better performance, particularly under shading. Micro inverters however tend to be expensive and they should be chosen using a proper cost-benefit analysis.

- **String Combiner and junction boxes**: these are used to connect different solar panels together in an array and route the DC electricity produced to the inverter.
- **Cables, breakers, switches and disconnects** are to appropriately chosen to ensure system safety and performance. Rooftop cables are typically exposed to the environment, and should therefore be able to withstand UV light, ozone, heat and rain or hail without degrading. Cables used in PV installations are specifically manufactured to be UV resistant.
- **Batteries**: used for storing the energy generated from solar PV in stand-alone systems. Batteries are not necessary for grid-connected solar PV systems.
- **Charge controllers**: charge controllers are used to protect the battery and again mostly used in stand-alone systems along with the battery.
- **Meters**: it is important to measure the energy produced by the solar panels, so that the building energy use can be matched and the performance of the solar PV systems can be monitored.
- **Monitoring systems**: apart from meters, there are other parameters that can be monitored to understand and fine-tune the operations of the Solar PV system. Weather stations measure and track temperature, insolation and wind speeds, so that the performance of the system can be compared to what should have been generated, given the conditions. Pyranometers or reference cells, anemometers and thermometers are all common in PV installations.

## *7.1.2 Building Integrated PhotoVoltaic (BIPV)*

PV modules can be incorporated into the building's façade or other building envelope elements. In this way, they serve a dual purpose of producing electricity and enhancing the building's aesthetic features by replacing traditional building envelope materials. BIPV can help maximize the generation of energy from solar PV in the building, while be more aesthetically pleasing in integrated well with the right type and the right location. Another advantage of integrated photovoltaics is that the initial cost can be offset by reducing the amount spent on building materials and labour that would normally be used to construct the part of the building that the BIPV modules replace.

BIPV can be integrated on opaque surfaces and in this case, the traditional solar panels can also be used. However, there are now semi-transparent or translucent solar PV modules that can be used on glazing, skylights and other building envelope locations where completely opaque modules would not work (e.g. canopies and plant shelters). Thin films as well as organic and dye-sensitized solar cells are increasingly being used in BIPV applications. Figure 7.2 shows a photo of BIPV modules installed at the Geneva international airport.

Just as in the case of the rooftop PV systems, the balance of system (BOS) components are still required for BIPV installations. The routing of cables and wires, and placement of the BOS equipment is more challenging in case of BIPV. In rooftop systems, the BOS equipment can be located on the roof itself with

**Fig. 7.2** Photo of BIPV modules installed at Geneva Airport

appropriate weather proofing. BIPV systems also may suffer from poor operational performance as the orientation of the PV modules may not be in line with the sun path and hence may result in sub-optimal production yields. It is rare to find a BIPV system that produces more energy than the rooftop solar PV system.

## 7.2 Solar Thermal Collectors

Apart from sunlight, the heat from the sun can also be used to generate energy in form of hot water or air that be used directly in the building. The main use of this technology is in residential buildings where the demand for hot water has a large impact on energy bills. This generally means a situation with a large family, or a situation in which the hot water demand is excessive due to frequent laundry washing. Commercial applications include laundromats, car washes, military laundry facilities and eating establishments. The technology can also be used for space heating in buildings.

The solar collector is the key component of the solar thermal system that collects the solar radiation and transfer the heat to the heating medium. Three different type of solar collectors are used in water heating systems viz. flat-plate collectors, unglazed collectors and evacuated tube collectors.

### *7.2.1 Flat Plate Collectors*

Flat plate collectors have a large heat absorbing area or an absorber plate. The absorber plate is made of copper, steel or plastic and the top surface of the plate is coated with high absorptance black paint or other selective coatings that have high absorptance for solar radiation and low emittance for long wavelength radiations. Flow passages for the heating medium are formed by bonding pipes, tubes or coils to the absorber plate. The assembly of absorber plate and tubes is enclosed in a casing and covered with a glazing such as toughened glass or plastic. These glazing create a 'green-house gas' effect by trapping the solar radiation effectively by having very low transmittance for long wavelength radiation that is reflected from the absorber. The key components of flat plate solar collectors are shown in Fig. 7.3.

Flat plate collectors can reach temperatures of about 80–100 °C for the heating medium, which is typically water. In case of direct air heating, air typically passes along the front or back of the absorber plate while scrubbing heat directly from it. Heated air can then be distributed directly for applications such as space heating and drying.

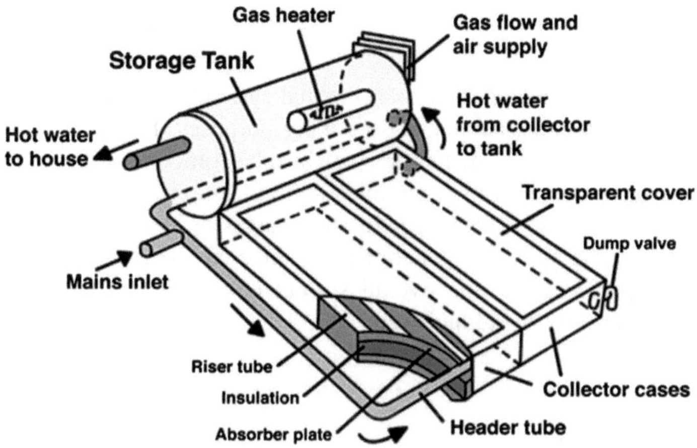


**Fig. 7.3** Flat plate solar hot water collectors—an illustration (yourhome.gov.au)

### *7.2.2 Unglazed Collectors*

Unglazed collectors may provide water at temperatures that are 10–15 °C higher than the ambient air temperature. these collectors are manufactured by extrusion from polymers such as polypropylene, polyolefins, polycarbonate, and other materials with UV Stabilizers. The collectors are used without insulation or glazing in order to facilitate ease of fabrication and lower the costs of manufacturing. A common application for such collectors is in heating water in swimming pools.

### *7.2.3 Evacuated Tube Collectors*

One of the main problems with flat-plate collectors is the heat loss due to conduction and convection to outside air and the degradation of the absorber material and glazing due to air contact. Evacuated tube collectors are composed of multiple evacuated glass tubes each containing an absorber plate fused to a heat pipe. The heat is transferred to the transfer fluid of a domestic hot water or hydronic space heating system in a heat exchanger called a "manifold". The manifold is wrapped in insulation and covered by a protective sheet metal or plastic case.

The vacuum that surrounds the outside of the tube greatly reduces convection and conduction heat loss, therefore achieving greater efficiency than flat-plate collectors, especially in colder conditions. The reflective coating for the design is encapsulated in the vacuum inside of the tube and effectively prevents the heat loss due to reflected radiation. Evacuated tube collectors can be used to deliver low to medium temperature heating from 90–150 °C at reasonable efficiencies.

### 7.2.4 Solar Thermal Based Cooling and Air-Conditioning

In most cases, the hot water from solar thermal system can be directly used within the building for shower, laundry and cooking purposes. However, the hot water generated from the solar thermal system can also be combined with thermal chillers such as absorption chillers and desiccant dehumidification applications as discussed in Chap. 5 (Sects. 5.1.4 and 5.1.5). In such cases the design often incorporates storage tanks for the hot water generated and since the thermal chillers have a low coefficient of performance, the overall system efficiencies in terms of the used electric power need to be studied well.

## 7.3 Wind Energy

Wind energy harnesses the energy available in naturally flowing air at high wind speeds. The key component of wind energy systems is the wind turbine, that convert the mechanical motion driven by high speed wind (or kinetic energy) into electricity. Wind speed varies roughly by the square root of the change in of altitude, so the top of buildings will experience faster wind than the ground level. This is also the reason that a typical wind turbine is elevated on a tall tower.

Wind turbine design falls into two basic configurations based on the axis of rotation of the turbine blades, viz. vertical axis wind turbines and horizontal axis wind turbines. Horizontal axis means the rotating axis of the wind turbine is horizontal, or parallel with the ground. Most commercial large wind farms would only use horizontal axis wind turbines as they are able to produce the most electricity at steady wind conditions. However, they are generally large structures, heavier and do not produce well in turbulent wind conditions. Hence for most building integrated wind turbine applications, the vertical axis wind turbines are preferred option. With vertical axis wind turbines, the rotational axis of the turbine stands vertical or perpendicular to the ground. Vertical axis turbines are powered by wind coming from all 360°, and even when the wind is blowing in different direction. The output energy of the vertical axis wind turbines is however significantly lower. Figure 7.4 shows a photo of vertical axis wind turbines installed on a building at the Hong Kong University campus.

In case of buildings, due to space constraints and the fact that airflow can be turbulent and affected by several obstructions (such a neighbouring buildings), the vertical axis wind turbines would be a better choice. wind turbines may generate noise and vibration and hence their location and mounting has to be chosen carefully. Large scale vertical axis wind turbines could be used in some rare cases next to buildings in open fields and the electricity produces could be wired into the building for use.

**Fig. 7.4** Photo of vertical axis wind turbines installed on top of a building at the Hong Kong University campus

## 7.4 Geothermal Energy

Geothermal energy is the heat from the earth, below its crust. Although geothermal heat from deep below the earth's crust is now used for electricity production, it is difficult to integrate it in buildings. The shallow ground, the upper 10 feet of the earth, maintains a nearly constant temperature. Like a cave, this ground temperature is warmer than the air above it in the winter and cooler than the air in the summer. This temperature differential can be harnessed in geothermal heat pumps to cool or heat buildings.

Geothermal heat pump (GHP) systems consist of basically three main parts: the ground heat exchanger, the heat pump unit, and the air delivery system or ductwork. Heat pumps have been described earlier in Chap. 5 (Sect. 5.1.3). The heat exchanger is basically a system of pipes or tubes called a loop, which is buried in the shallow ground near the building. A fluid (usually water or a mixture of water and antifreeze) circulates through the pipes to absorb or dissipate heat within the ground. Figure 7.5 shows and illustration of the underground piping layout.

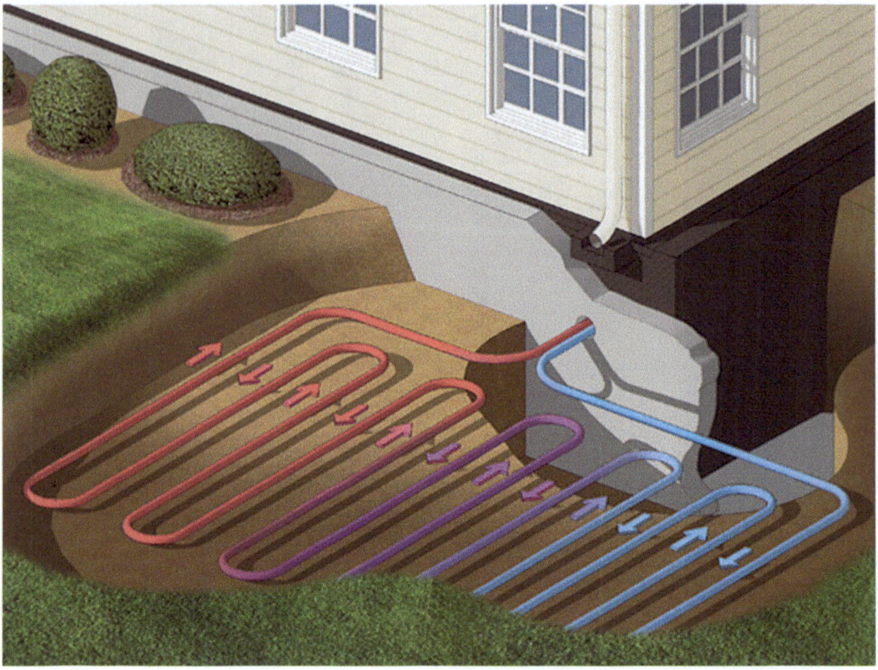

**Fig. 7.5** The underground loop layout in geothermal heat pump systems (SWRE.com)

The loop used in a heat pump system can be classified in two broad types viz. closed-loop system and open-loop system. Most closed-loop geothermal heat pumps circulate an antifreeze solution through a closed loop tubing system that is usually made of plastic tubing that is buried in the ground. A heat exchanger transfers heat between the refrigerant in the heat pump and the antifreeze solution in the closed loop. The open-loop system uses well or surface body water as the heat exchange fluid that circulates directly through the GHP system. Once it has circulated through the system, the water returns to the ground through the well, a recharge well, or surface discharge. This option is obviously practical only where there is an adequate supply of relatively clean water, and all local codes and regulations regarding groundwater discharge are met.

## 7.5 Energy Storage

One of the solutions to ensure smooth operation of the building with high amount of renewable energy integration is energy storage. In some cases, extra electricity is generated when the sun is shining brightly or when turbines receive sustained gusts of wind throughout the day. In other cases, on the contrary, sunlight concealed by cloudy days and fluctuating breezes limit the usefulness of renewable generation through solar and wind energy options respectively. energy storage can provide a solution in both cases by storing energy and reutilising it at the right time. Energy storage can help stabilize the grid, and avoid spikes in electricity use that are not only expensive but may lead to blackouts or brownouts. Some building owners may even opt to be fully grid independent or off-grid using energy storage.

Energy can be stored in electrical, mechanical, electro-chemical, chemical and thermal means, while delivering the final energy in electrical or thermal form. There are various technologies available today for energy storage such as thermal storage, batteries (electro-chemical storage), flywheel (kinetic energy), pumped hydro (mechanical), compressed air (mechanical), hydrogen cycle (chemical storage). Not all of these energy storage alternatives are possible to be integrated in buildings and the most popular options used in some buildings today are thermal energy storage and batteries.

### 7.5.1 Thermal Energy Storage

As heating and cooling takes up a large portion of the building's energy use, storing energy in the form that it will eventually be used makes sense. Storage mediums such as ice are very effective for storing and releasing large amounts of cooling with the least possible waste, making ice storage a good solution. This is carried out by augmenting a chiller-based cooling system with on-site thermal storage tanks that form and store ice. In some cases, chilled water can be stored instead of ice and there are other phase-change materials such as molten salt (for heating application) that are under development to be used as thermal storage medium. Often times, space is the main challenge for thermal storage technology implementation. There are also thermal storage technologies via chemical reactions such as adsorption that offer higher energy densities and thereby reducing the space for storage (IRENA 2013).

### 7.5.2 Battery Storage

Batteries offer commercially viable energy storage options in many cases, especially when the electric grid supply is not reliable. Batteries are good choice for storing the renewable electricity generated by Solar PV or Wind power systems. they can offer the necessary buffer capacity of supply for intermittent renewable energy sources. The most common type of battery storage technologies are:

- Lead-acid batteries: one of the oldest and most developed battery technologies that is now typically available cheaper than any other battery option. They have short cycle-life (2–4 years typically) and long recharge times.

- Nickel-based batteries: There are two types of nickel batteries, the older, nickel-cadmium (NiCd) batteries, and the newer, nickel metal-hydride (NiMH) batteries. NiCd are relatively inexpensive, able to sustain deep discharge, recharge quickly, and have a long cycle life. However, NiCd are extremely environmentally unfriendly because of the use of toxic cadmium and are been phased out in favour of NiMH batteries that in comparison also have a better energy and power capacity.
- Lithium-ion batteries: The first generation of such batteries allowed storing more than twice the energy compared to nickel or lead batteries of the same size and mass. Today, the lithium-ion batteries offer the promise of high energy, high power, high efficiency, longer life, and easier state-of-charge control at lower weight, volume, and reasonable cost.
- Redox flow batteries: Vanadium redox-flow battery (VRB) is one of the mostly studied rechargeable flow batteries. The higher scale capacities possible from vanadium redox batteries make them well-suited to use in large power storage applications. The main disadvantages with vanadium redox technology are a relatively poor energy-to-volume ratio, and the system complexity, in comparison with standard storage batteries.

**Fig. 7.6** Photo of a Vanadium Redox-flow battery (VRB) installed in the basement of building in Singapore

The above list of battery technologies is not exhaustive and new technologies in this field are constantly evolving. Batteries are commonly used for applications such as back-up power for critical building service devices such as Building Management System (BMS) and for safety back-up in Elevators. The initial higher costs of batteries and their relatively shorter life-spans are the major factors against their large scale adoption in buildings for larger energy storage applications. There are however many buildings and households in the world that have started adopting batteries to be able to integrate renewable energy and reduce their dependence on the electric grid supply. Figure 7.6 shows a photo of a VRB storage system used in a building as a test-bed for renewable energy and electrical vehicle integration.

## 7.6 Smart Energy Management Systems

With a variety of renewable energy options in buildings, the energy management aspect of integrating these generation sources into the energy consumption mix of buildings is important. Most of the renewable energy options that directly produce electricity (solar PV and wind energy) tend to be intermittent and heavily reliant on the weather conditions. For example, in cloudy conditions, the energy generation from rooftop solar PV can drop significantly. Also buildings nowadays operate in the era of a smart grid where the electricity grid options are diverse and managed much more dynamically.

Specifically, the main interaction between the building and the electricity grid is the commercial grid connection that is contractually controlled. In simple terms, the electricity utility company sells the grid power to the building at a pre-determined contractual rates or electricity tariff. In many countries, the purchase of electricity from the utility supplier is liberalised and consumer has various options for contracting their electricity demand with a supplier. The electricity tariff can be based on the time-of-use (TOU) pricing principle and the building would also be allowed to export electricity generated from the integrated renewable energy sources to the grid. A smart grid mechanism known as 'demand response' also allows for the consumers to get paid for reducing their power demand at certain times of the day.

To facilitate these interactions with the electricity grid, the building can benefit from a smart energy management system. A smart energy management system (SEMS) is essentially a decision making tool to ensure that the energy resources and demand is actively managed considering the interaction of the building with the grid. Various energy generation sources and possibly also energy storage devices such as batteries can be integrated into the SEMS to make some key decisions to reduce the overall cost of electricity or maximise energy generation from renewable energy sources.

## References

BNEF (Bloomberg New Energy Finance) (2016) New energy outlook 2015—powering a changing world. Available online at: http://www.bloomberg.com/company/new-energy-outlook/. Accessed 2 Mar 2016

GBPN (Global Building Performance Network) (2016) Defining positive energy buildings—a spectrum approach. Available online from: http://www.gbpn.org/laboratory/positive-energy-buildings. Accessed 15 Mar 2016

IRENA (International Renewable Energy Agency) (2013) Thermal energy storage—technology brief, January 2013. Available online at: http://www.irena.org/Publications/. Retrieved 30 Dec 2015

SWRE (Touchstone Energy cooperative) (2015) Geothermal, Available online at: http://swre.com/content/geothermal. Accessed 25 June 2015

YourHome, Australia's guide to environmentally sustainable homes, Available online at: http://www.yourhome.gov.au/energy/hot-water-service. Updated 2013. Accessed 24 June 2015

# Chapter 8
# Water and Waste Management Technologies

Apart from energy, the usage of water and management of waste are important aspects of green buildings. Buildings use significant amount of water for various needs such as sanitation, kitchen, laundry, utilities, heating/cooling and landscaping needs. The availability of clean water is a major challenge in several parts of the world and hence water conservation in buildings is imperative. Also buildings can lead to significant waste generation throughout its lifetime, right from its construction phase to its demolition. The waste generated by the building occupants and facilities also needs to be managed in order to reduce its environmental impacts.

## 8.1 Water Consumption, Monitoring and Leak Detection

In buildings, water is typically consumed for the following purposes:

- Sanitation: toilets, showers and was basins
- Washing: laundry, cleaning of floors and other building surfaces
- Kitchen: food preparation, washing and drinking water
- Equipment: Heating, ventilation and cooling systems use water
- Landscaping: for maintaining greenery, gardens and plants in the building.

Typically water is imported into the building from the water utility provisions and then stored and transported to various places in the building through pipes and pumping system. There could be however different grades of water used for the different purposes that are described above and it's important to specify and understand the different water types and then devise the appropriate conservation measures.



N.Y. Jadhav, *Green and Smart Buildings*, Green Energy and Technology,
DOI 10.1007/978-981-10-1002-6_8

### 8.1.1 Identifying the Different Water Types in Buildings

#### 8.1.1.1 Potable Water

Potable water is water which fit for consumption by humans and other animals. It is also called drinking water, in a reference to its intended use. In many first world and developed countries the fresh water piped into buildings is potable water. In many other countries the water supplied by the utility to the building has to be further treated and purified to be fit for human consumption.

#### 8.1.1.2 Black Water

The term 'black water' is used to describe wastewater containing contaminants such as human faeces, urine etc. from toilet flushing and cleaning. It often end up in sewage for discharging out of the building almost immediately after use as it can pose health hazard for the occupants of the building.

#### 8.1.1.3 Greywater

Greywater refers to untreated wastewater that has not come into contact with sewage. It can be the wastewater generated from household uses like bathing and washing clothes, foods and dishware but not from toilets. In many utility systems around the world, greywater is combined with black water in a single domestic wastewater stream. Greywater can be of far higher quality than black water because of its low level of contamination and greater potential for reuse (Allen et al. 2010).

#### 8.1.1.4 Process Water

The term 'process water' is used typically used in relation to industrial plants, industrial processes, production facilities and laboratories. However in commercial buildings, the process of heating, ventilation and cooling also uses water and this can be classified as process water in buildings. The water used for air heating and cooling is often circulated in a closed loop and its temperature differential is used for air cooling. In case of heating this is referred to as the hot water or steam circulation and in case of cooling it is often referred to as chilled water circulation. In cooling towers, water is used for evaporative cooling and it is lot with the air as humidity. In a reverse way, when air is dehumidified, there is water generated from the air known as condensate water. Condensate water is essentially distilled water, low in mineral content, but may contain bacteria from the air.

### 8.1.2 Water Sub-metering and Leak Detection

Typically, the amount of water supplied to the building by the utility provider is based on a utility meter and its used to charge the building for its water usage. However, it's also important to understand the water usage within the building and its various provisions and this is where sub-metering helps. Sub-metering can provide important information to analyse water usage in a building and also provide insights into ways to reduce the wastage of water through benchmarking and other calculations such as water balances.

Leaks in water transport system (pipes, pumping and storage tanks) can be major cause of water wastage and loss in buildings as the undetected leak can keep on siphoning useful water (such as potable and process water) into sewage system. Numerous factors can cause leaks in buildings such as old or poorly constructed pipelines, inadequate corrosion protection in the piping system, poorly maintained valves, poor sealing and mechanical damage. In commercial and institutional buildings, water supplied by the main is often stored in tanks prior to its use. The structural stability of storage tanks can deteriorate over time due to various reasons leading to leaks. Leaks and water loss can also result from malfunction of equipment and instruments such as overflow control valves, shut-off valves, bypass valves and nozzles. A common causes of water leakage in many places are also the dripping taps, faucets and shower heads.

By conducting regular checks for leaks and routine maintenance of equipment, considerable amounts of water can potentially be saved in buildings. A water balance calculation and monitoring of water level in tanks are often simple ways to identify leaks. In commercial and institutional buildings, more complex measures such as continuous monitoring, overnight monitoring, and dynamic water balances may need to be used to determine the extent of leakage.

Not all leaks are however visible and hence can go undetected for a long time and can lead to major water losses. There are various methods for detecting water distribution system leaks. These methods usually involve using sonic leak-detection equipment, which identifies the sound of water escaping a pipe. These devices can include pinpoint listening devices that make contact with valves and hydrants, and geophones that listen directly on the ground. In addition, correlator devices can listen at two points simultaneously to pinpoint the exact location of a leak (EPD 2007).

## 8.2 Water Efficient Fittings

The best way to directly reduce water consumption in buildings is by using water efficient fittings that can results in immediate and measurable water savings.

### *8.2.1 Water Efficient Faucets and Tap Adaptors*

Aerators are devices that mix water and air and can reduce both water flow rates and splashing while increasing areas of coverage and wetting efficiency. Modern faucets come with integrated aerators and aerators as tap adaptors are also separately available that can be integrated with existing taps. Such systems can restrict water flow and result in significant water savings of 40–80 % (DrainWorks 2014). Figure 8.1 show a photo of commercially available aerator fixture for water faucets.

### *8.2.2 Water Efficient Shower Heads*

Similar to faucet aerators, efficient shower heads operate by mixing water flow with an air jet. These units provide satisfactory contact with water and achieve effective rinsing with much less water. These can cut water usage by 25–60 % (Drain Works 2014).

### *8.2.3 High Pressure or Trigger Spray Nozzles*

High pressure spray nozzles for cleaning of floors as well as dishes and utensils in kitchens can result in significant water savings. Such systems use the jet force of water for cleaning instead of relying on too much water to wash off the contaminants such as dirt and food particles stuck on utensils.

**Fig. 8.1** Photo showing an aerator fixture at a water faucet (DrainWorks 2014)

### 8.2.4 Automatic Water Shut-off Showers and Faucets

Showers and faucets with shut-off systems automatically cut the water flow once a predetermined amount of water has been used. Such systems require user input such as pressing the knob to re-activate the water flow. This can significantly reduce the water usage by avoiding wastage when the user forgets to turn the tap off after use. Hence in common toilets and bathrooms such as in sport facilities, schools, offices and dormitories, such system can save a lot of water without relying user initiative to only use as much water as required.

### 8.2.5 Water On-Demand Sensors

With the use of infrared or motion sensors that can detect a hand under a tap for example, the water flow can be regulated significantly. Such sensors are coupled with water cut-off or cut-in controllers that regulate water flow based on demand rather than human action such as turning a tap valve or pressing a knob to activate water flow. It is essential that such units have a quick response time in order to avoid user dissatisfaction. Figure 8.2 shows a photo of a bathroom with sensor controlled water taps.

**Fig. 8.2** Photo showing a sensor controlled water tap

### 8.2.6 Instant Water Heaters

Instant water heaters help reduce water consumption by avoiding the wait for hot water, which typically causes a lot of cold water to go down the drain. With the use of easily adjustable water mixers with temperature indicators, desired water temperatures can be more easily achieved, thereby wasting less water.

### 8.2.7 Low Volume and Dual-Flush Cisterns for Toilets

The low volume and dual-flush toilets allow for the use of a reasonably low amount of water to flush compared to the old full-flushed toilets. In dual flush cisterns, two options are available for flushing the toilet bowl and the user can use the low volume flush option to save water by using significantly less volume of water for flushing. This choice is typically based on user need e.g. the low flush option can be used after urinating and the high volume flush can be saved for flushing faeces after defecation. Over a long period of time, such systems can result in significant water savings in buildings.

However, these dual flush systems usually require the replacement of not only the cistern and the flushing mechanism, but also the toilet bowl. Therefore, they should be considered when replacing old models or installing new toilets. A simple option to reduce the flush volume in older toilets is to is to place a displacement object inside the toilet cistern. These objects sit inside the cistern tank permanently occupying a reasonable volume without interfering with the operational mechanism of the flush system. Another possible way is to use the so-called toilet dams, which are barriers placed inside the cistern, creating dry compartments and thereby reducing the amount of water used in each flush.

### 8.2.8 Water-Less Urinals

Waterless urinals otherwise known as flush-less urinals came into existence in the bid to create an efficient water usage. Klaus Reichardt, a German, invented the waterless urinals, which have a device installed in it with a sealant liquid such as oil that allow urine to enter the waste water system through one-way or liquid sealed valves. Liquid sealed valves allow urine to seep through the oil blocking the outlet, making use of the difference in densities between the two fluids. Figure 8.3 illustrates the mechanism of water-less urinals.

It is estimated that more than 160 billion gallons of water are flushed down the drains each year due to urinal usage (ZME Science 2013). The use of waterless urinals can hence significantly result in water conservation in buildings. It's

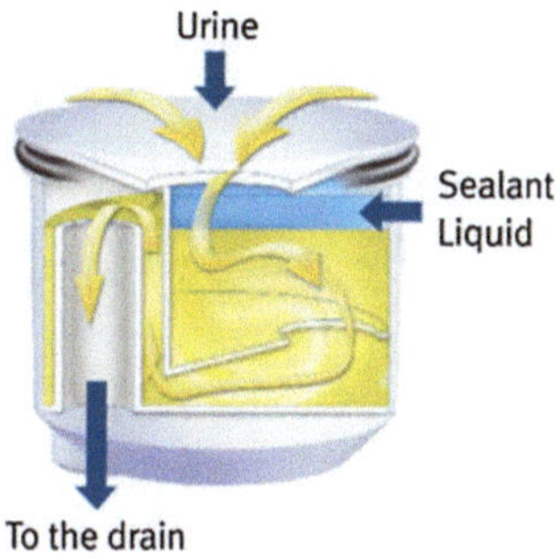


**Fig. 8.3** Mechanism of a water-less urinal (ZME Science 2013)

important however to choose the right waterless urinal and clean it on a regular basis (most urinals can be cleaned simply by wiping the bowl with an all-purpose cleaner) to ensure that odour and hygiene issues are controlled well.

### *8.2.9 Composting Toilets*

A composting toilet uses a predominantly aerobic processing system to treat human excreta, by composting or managed aerobic decomposition. The human excreta is usually mixed with sawdust, coconut coir or peat moss to facilitate aerobic processing, liquid absorption, and odour mitigation. Composting toilets produce a compost that may be used for horticultural or agricultural soil enrichment if the local regulations allow this. The composting chamber can be constructed above or below ground level. It can be inside a structure or include a separate superstructure. These toilets generally use little to no water and may be used as an alternative to flush toilets. Figure 8.4 illustrates the concept of a typical composting toilet.

### *8.2.10 Water Efficient Equipment Labels*

The Environmental Protection Agency (EPA) in the United States of America (USA) has a programme called EPA WaterSense that classifies and labels water efficient equipment and products. The consumers can thus choose more water efficient equipment by simply looking at the WaterSense logo on the products such as showerheads, toilets, bathroom sink faucets, urinals, etc. Products and services that have earned the WaterSense label have been certified to be at least 20 % more efficient without sacrificing performance (EPA 2016).

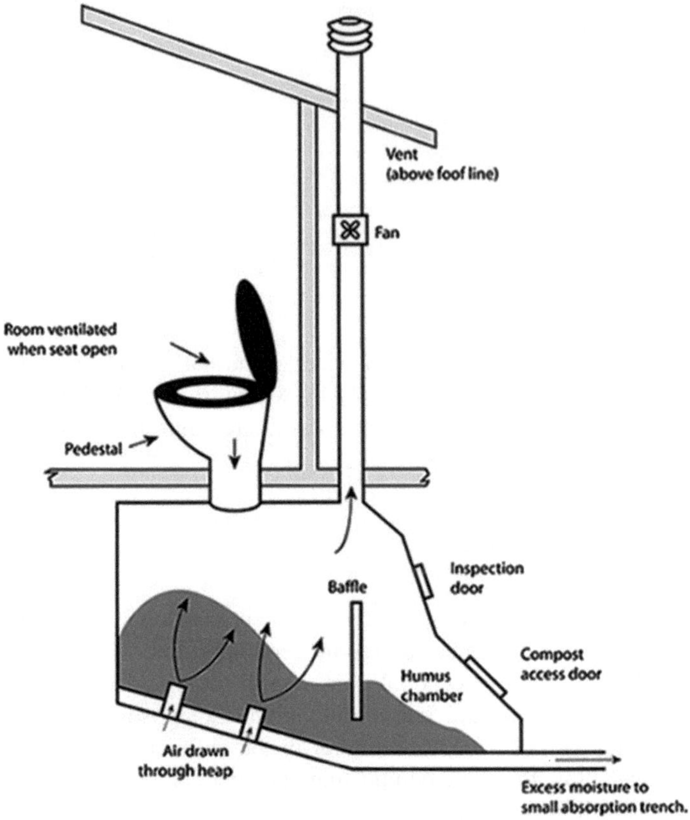


**Fig. 8.4** Diagram illustrating a typical composting toilet (*Source* WaterNSW)

## 8.3 Greywater Recycling

Earlier we discussed greywater as any domestic, household or office-building wastewater without toilet water contamination i.e. different from black water from toilet flushing that is send to sewage immediately. Recycling of greywater has become a matter of importance to the survival of freshwater as the supply of the available freshwater cannot meet the demand of its users.

Most greywater is easier to treat and recycle than black water, because of lower levels of contaminants. If collected using a separate plumbing system from black water, domestic greywater can be recycled directly within the home, garden or company and used either immediately or processed and stored. If stored, it must be used within a very short time or it will begin to decompose due to the organic solids in the water. Recycled greywater of this kind is never safe to drink, but some treatment steps can be used to provide water for washing, gardening, irrigation or flushing toilets.

There are three major ways of treating or recycling greywater namely:

1. Diversion system: Before immediate re-use, this system may filter and disinfect but it doesn't allow for storage. The reuse of greywater for toilet flushing and garden irrigation has an estimated potential to reduce domestic water consumption by up to 50 % (Maimon et al. 2010). Some smart designs integrate the hand-wash sink with the flush water tank of the toilet to enable direct re-use of greywater (see Fig. 8.5).
2. Physical and chemical greywater treatment: This system allows filtered, disinfected and treated greywater to be stored for future use. The treatment destroys the bacteria and other microorganisms that can grow in numbers in immoveable water. Chlorine, ozone or ultraviolet light are all available and efficient disinfection methods. Meanwhile, the disadvantage is that Chlorine and ozone have the tendency to produce toxic by-products, ozone and ultraviolet can be vastly affected by various organic elements in greywater.
3. Biological greywater treatment system: This system uses natural water processing technologies most especially aeration and membrane bioreactors. Through the aeration biological treatment, air is bubbled to move oxygen from the air into the greywater. However, the bacteria in it takes in the dissolved oxygen in the process they digest the contaminated organics thereby decreasing the concentration of pathogens. Moreover, this method can handle varieties of greywater regarding their qualities and quantities which allow for indefinite storage of the recycled water. This system requires high operating cost.

**Fig. 8.5** Toilet that re-uses the greywater from the sink directly above it (Caroma 2016)

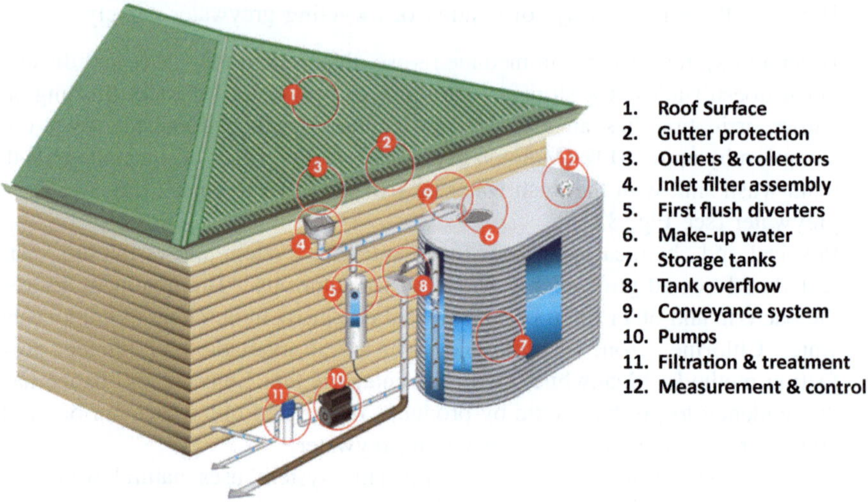


**Fig. 8.6** Components of a rainwater harvesting system in buildings (Rainwater Harvesting Supply Company 2016)

## 8.4 Rainwater Harvesting

Rainwater harvesting is a technique used for collecting, storing and using rainwater. The rainwater is collected from various surfaces such as rooftops and other man-made aboveground hard surfaces. It is then stored and treated for eventual use in the building. Harvested rainwater can be used for varieties of purposes such as watering of the garden, irrigation, household use with deep treatment, and livestock, etc. It is drinkable and it retains its pureness even after the long term of storage. Groundwater recharge can be carried out through harvested rainwater as well. Figure 8.6 shows the components of a rainwater harvesting system in buildings.

## 8.5 Water Efficient Landscaping

Green buildings have greenery incorporated in buildings and the plants and vegetation used in landscaping can consume significant amount of water. There is significant potential for reducing water used for landscaping by choosing the right plant species and irrigation techniques.

### *8.5.1 Drip Irrigation*

Drip irrigation is an irrigation system that supplies waters directly and slowly to the roots or onto the soil surface of numerous plants in a regulated and controlled

**Fig. 8.7** Photo of a drip irrigation tubing layout in a building landscaping

manner using valves, pipes, tubing and emitters cohesively. This can save water as irrigation is done is regulated manner and prevents evaporation losses. Figure 8.7 shows photo of a landscaping plants in a building with drip irrigation provisions.

### 8.5.2 Rain Sensors to Control Irrigation

The rain sensor irrigation system encourages the moderate and wise use of water by leveraging on natural rain water whenever and wherever possible. A rain sensor typically works for the water sprinkler irrigation system. The sprinkler rain sensor operates through a gauge installed to the fence in proximity to the lawn or garden incorporated with the sprinkler system. There is a disc inside the gauge, it takes in water and expands as the rain continuously falls. These send a signal to the sprinkler system regulator, which stops the electronic signal that turns on the sprinklers until the disc contracts to their standard size. However, with this, the sprinkler regulator knows it's time to start the spraying and it turns on the sprinkler system automatically once again.

### 8.5.3 Xeriscaping

Xeriscaping is a technique that emphasizes the selection of plants for water conservation. Drought-tolerant plants are an essential part of water efficient landscapes.

They are adapted to water scarce environments and therefore require minimal supplemental irrigation. They also require less maintenance than their water-needy counterparts. The specific plants used in xeriscaping depend upon the climate.

## 8.6 Water Reduction in Cooling Towers

Cooling towers are supplementary cooling systems used in water cooled chiller plants as described earlier in Sect. 5.1.1. Cooling towers rely on evaporative cooling to reduce the water temperature in order to reject heat in the condenser loop when producing chilled water. Although cooling towers are efficient heat rejection devices, they can consume a significant amount of water in buildings.

Cooling towers consume water in the following ways:

(a) Evaporation of water: this is a necessary thing for the system operation and the water savings can only be achieved in this area by optimising the system operations and reducing the cooling load.
(b) Bleed: Due to the evaporation of water, the cooling tower concentrates and accumulates both dissolved and undissolved or suspended solids that can lead to fouling of the system. This can be controlled by bleeding a portion of the circulating water from the system and replacing it with relatively clean make-up water.
(c) Drift: it is the small water droplets that leave the tower entrained in the tower discharge air. It is avoided by installing drift eliminators in the cooling tower. Controlling fan speeds and preventing ambient wind speeds from impacting the tower are ways of further reducing potential drift losses.
(d) Windage and splash out: Splash out refers to water leaving the tower via the air intakes and other openings in the tower casing. It can be reduced by installing anti splash louvers on the tower air intakes.
(e) Overflow and Leaks: Overflow is an uncontrolled water loss caused by water flowing back into the cold water basin and can be caused by poor pipe work, design and installation. Leaks result in water and chemical wastage and disturb the balance of water treatment systems by diluting the system with more make-up water than expected. Leaks are minimised by periodically reviewing the cooling tower and associated system for leaks and checks through establishment of a water balance as shown in Fig. 8.8.

### *8.6.1 Improving Cycles of Concentration*

The term cycles of concentration or concentration ratio refers to the ratio of impurities or the total dissolved solids (TDS) in the circulating water to the TDS in the make-up water. The selection of an appropriate level of cycles of concentration

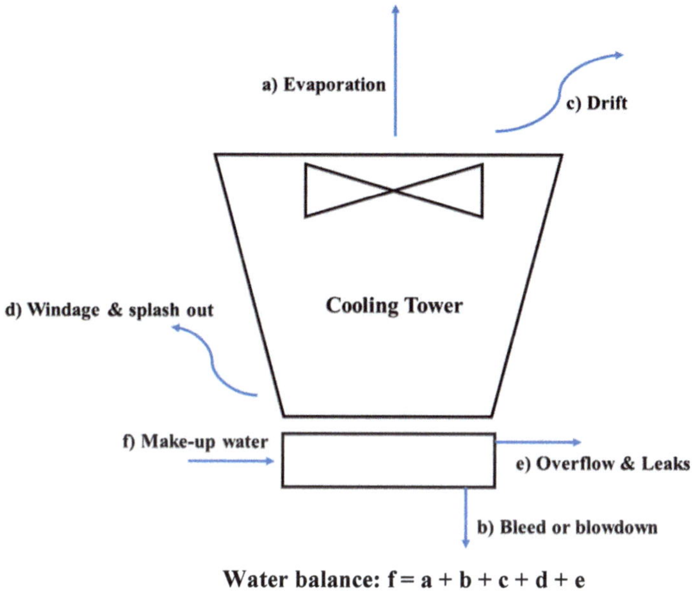


**Fig. 8.8** Water balance in a cooling tower (adapted from AIRAH 2009)

is a complex process and operators and water treatment service providers need to adopt a holistic approach to these considerations. The upper limit to the number of cycles of concentration that can be achieved is primarily determined by the purity of the make-up water. By increasing the cycles, this reduces the bleed, thereby reducing the amount of make-up water required by the system. Water treatment options can be used to assist in efforts to increase allowable cycles of concentration within the system. The make-up water can be pre-treated or filtered to improve its quality and hence allowing an increase in concentration ratio, so reducing bleed. Table 8.1 shows the potential reduction in water by increasing the cycles of concentration.

## 8.7 Reducing Construction Waste

The construction industry is one of the major waste producers worldwide and in the United Kingdom (UK) alone, the construction industry is responsible for some 120 million tonnes of construction, demolition and excavation waste every year, which is around one third of all waste arising in the UK (WRAP 2016). Figure 8.9 shows a photo of typical construction waste found on building construction sites.

**Table 8.1** Potential water savings that can be achieved by increasing the cycles of concentration in a cooling tower (AIRAH 2009)

| Percentage of make-up water saved | | | | | | | | | | | |
|---|---|---|---|---|---|---|---|---|---|---|---|
| Old cycles of concentration number C1 | New cycles of concentration number C2 | | | | | | | | | | |
| | 2.0 | 2.5 | 3.0 | 3.5 | 4.0 | 5.0 | 6.0 | 7.0 | 8.0 | 9.0 | 10 |
| 1.5 | 33 % | 44 % | 50 % | 53 % | 56 % | 58 % | 60 % | 61 % | 62 % | 63 % | 64 % |
| 2.0 | | 17 % | 25 % | 30 % | 33 % | 38 % | 40 % | 42 % | 43 % | 44 | 45 % |
| 2.5 | | | 10 % | 16 % | 20 % | 25 % | 28 % | 30 % | 31 % | 33 % | 34 % |
| 3.0 | | | | 7 % | 11 % | 17 % | 20 % | 22 % | 24 % | 25 % | 26 % |
| 3.5 | | | | | 5 % | 11 % | 14 % | 17 % | 18 % | 20 % | 21 % |
| 4.0 | | | | | | 6 % | 10 % | 13 % | 14 % | 16 % | 17 % |
| 5.0 | | | | | | | 4 % | 7 % | 9 % | 10 % | 11 % |
| 6.0 | | | | | | | | 3 % | 5 % | 6 % | 7 % |

**Fig. 8.9** Photo of construction waste on a building construction site (WRAP 2016)

### *8.7.1 Reuse and Recovery of Construction Materials*

The reuse of material components and/or entire buildings has considerable potential to reduce the key environmental burdens (e.g. embodied energy, $CO_2$, waste etc.,) resulting from construction. Reuse of materials is possible with existing materials on site or use of new material for construction with high recycled content. The best opportunity to think about reuse is during the design stage when the designer can make a conscious effort to identify and specify the reuse of materials from the same or other buildings sites and also specify the recycled content in new materials being used on site. The other opportunity is for materials to be recovered effectively during the life of the building when maintenance and refurbishment is undertaken or when the building comes to the end of its life.

The following are some techniques to maximize reuse and recovery of construction material:

1. Reuse of excavated material as drainage base and mound features in landscape design.
2. Reuse or recycle of tarmac and asphalt for constructing pathways, construction storage space and hard standing for plants.
3. Reuse of top soil for green roofs and other landscaping use.
4. Reuse of existing landscape items if possible after repairing and refurbishment if required, rather than completely discarding them.
5. Use of recycle aggregates in concrete mix as fill.
6. Use pf cement substitutes such as PFA (pulverized fuel ash) and GBBS (ground granulated blast-furnace slag).
7. Recycling concrete elements as aggregates and use them as a thermal mass for heat energy storage.

8. Reuse of packaging material for other purposes or return it to supplier/manufacturer.
9. Reuse of timber for cladding, fencing and other landscaping uses.
10. Timber from demolition sites can be de-nailed and used as wood chips for composting, landscaping top mulch or taken off-site for energy generation plants that use wood-chips as fuel.
11. Reuse of bricks and blocks for masonry, internal partitions and fair faced cladding, while slate can be used for roofing and landscaping.
12. Salvaging good quality doors and other auxiliary fittings and fixtures that can be set aside for reuse or donated to charity.
13. Reuse of existing buildings on site for contractor's site establishment and use of temporary site establishment buildings that can be reused.

### 8.7.2 Optimising Material Usage During Construction and Deconstruction

Designers also need to consider how work sequences affect the generation of construction waste and work with the contractor and other specialist subcontractors to understand and minimise these without compromising the design concept and safety. This can be done by solutions that use less material and simplify the structural solutions as much as possible. For example, using concrete solutions like post tensioning instead of cast in situ reinforced concrete and reusable/modular shuttering for slabs cores. Designers should also review foundation solutions to ascertain if options such as rotary or displacement piles can be used and excavation can be avoided by optimizing building position.

Another good practice is to carry out three-dimensional co-ordination exercises to eliminate all dimensional conflicts that lead to extensive on site waste. Mechanical, Electrical and Plumbing (MEP) layout and distribution routes can be planned to reduce builders works by consolidating risers, ducts and pipe distribution network. Standardizing windows, doors and glazing areas as well as light fitting and lamps can also result in good material supply and packaging efficiencies. Material planners could also specify materials from local sources and procurement routes that reduce packaging.

The building can be designed for deconstruction flexibility at an early stage in the design process. This can be done by using mechanical fixtures and even structural elements that can be easily dismantled. Use of design foundations that can be easily retracted from ground for reused should be considered. To facilitate easier dismantling of bricks and blocks, lime mortar or other mortar could be used instead of concrete filling. Use of gluing and composite materials can be avoided to enhance deconstruction flexibility.

### 8.7.3 *Pre-fabricated and Pre-finished Volumetric Construction (PPVC)*

The benefits of off-site factory production in the construction industry are well documented and include the potential to considerably reduce waste especially when factory manufactured elements and components are used extensively. Technologies used for off-site manufacture and prefabrication range from modern timber and light gauge steel framing systems, tunnel form concrete casting through to modular and volumetric forms of construction (WRAP 2016).

Prefabricated Prefinished Volumetric Construction (PPVC) is a construction method whereby free-standing volumetric modules that are complete with finishes for walls, floors and ceilings, are pre-fabricated or manufactured in an off-site fabrication facility. They are then installed in a building under building works almost in a 'LEGO-style'. PPVC can significantly speed up construction and can potentially achieve a productivity improvement of up to 50 % in terms of man-power and time savings. It can also reduce dust and noise pollution while enhancing construction site safety and material usage. PPVC has become a new industry standard and is even mandated for new construction activities. E.g. in Singapore, PPVC is mandatory for selected non-landed residential Government Land Sale sites from 1 Nov 2014 onwards (BCA 2016). Figure 8.10 shows photo of a PPVC construction site in Singapore.

## 8.8 Waste Reduction, Reuse and Recycling

Apart from construction waste, there is a lot of waste generated by people who occupy, use and manage buildings. From stationary to plastic to food, there are a lot of things that people consume in buildings and that end up in the waste-bin. Most of us are familiar with the golden rule of waste management or the 3R rule of Reduce, Recycle and Reuse. While these rules are for people to follow, building owners and facility managers can make their application easy by making certain provisions in the building such as recycling bins, multiple special purpose waste chutes and options to use less paper and plastic in buildings.

### 8.8.1 *Recycling Bins*

By providing recycling bins, apart from the general waste bins, building owners and facility managers can encourage people in the building to recycle more. The selection of the bins, their numbers, placement and type of waste to be recycled (paper, plastic, cans, glass, etc.) can be customised to the building type and its occupant profile. In today's world, electronic waste (or e-waste for short) is also a

**Fig. 8.10** Site photo of a PPVC assembly in Singapore

major problem and most of the electronic goods and appliances can be recycled for metal extraction and possible reuse after repair and refurbishment. Provision of e-waste bins for end-of-use electronic appliances and consumables such as batteries can significantly enhance the recycling rates and even avoid hazardous consequences of such waste ending up in landfills and incineration plants.

Some recent advances in waste recycling bin design have in-built compactors to be able to accommodate more waste and increase the productivity of waste management operations. The compacting operation can even be powered by use of sustainable energy such as solar photovoltaic that can harness sunlight whenever available (see Fig. 8.11 for a photo of such recycling bin).

## *8.8.2 Dual or Multiple Waste Chutes*

Waste chutes are common in communal housing buildings, where waste from several houses can be conveniently collected at a single location instead of door-to-door collection. However, this convenience is often abused by some occupants and all sort of waste gets thrown-in and sometimes in an inconsiderate manner. Building owners can consider incorporating dual or multiple waste chutes that are dedicated to certain waste disposal type. Provision of a separate waste chute for recyclable items makes it easier for occupants to dispose household waste for recycling. Some building owners provide separate waste chutes for organic and inorganic waste collection, so that organic food waste can be diverted to composting.

**Fig. 8.11** Photo of a solar-powered waste recycling bin installed outside a building in Singapore

### *8.8.3 Paper-Less Office*

Paper-less office is not a new concept, but it had been for most a dream that never comes true. It seems now increasingly more relevant in the era of handheld tablet devices, online or cloud storage, high storage capacity of removable storage devices (e.g. USB storage) and increase in data transfer speeds. Surprisingly, 50 % of the waste of businesses is composed of paper. Each year, the world produces more than 300 million tons of paper. According to the U.S. Environmental Protection Agency, printing and writing papers typically found in a school or office environment such as copier paper, computer printouts and notepads, comprise the largest category of paper product consumption. This is highly alarming in this digital era where electronic devices are so cheap and easily accessible.

Although a full paper-less office maybe a far reaching ideal, a near paper-less office is now possible with the available technologies and tools. The following tips can help in moving towards a nearly paper-less office:

1. Document Management system: hold reports and important documents online in a document management system that can help archive, store and share documents in a convenient way. There are now several document management tools and software available at an individual and enterprise level.
2. Adjust default settings for computers and printing.

   a. Set default paper-use in printers to double-sided printing and black-and-white prints.

   b. Set creation of PDF (portable document format) instead of paper printer as default for printing and provide necessary electronic storage space for storing PDF documents.
   c. Adjust computer settings to make on-screen reading easier and discourage printing of emails and other reference documents.

3. Reduce transactional paperwork through:
   (a) Paperless invoices, work orders and statements.
   (b) Electronic order processing and invoicing.
   (c) Online banking and money transfer (instead of written cheques).
   (d) Sharing electronic documents and using e-signature tools for contracts and other legal documents.
   (e) Internal online routing of documents for approval and endorsement.
   (f) Emailing documents instead of faxing.
4. Use digital technology for note-taking: using handheld devices such as tablet computers and smart phones for note-taking not only saves paper but also allows for easy retrieval and improves productivity.
5. Recycle and Reuse paper. Most copiers and printers now allow printing on reused paper through a special tray that can be selected for printing on such paper. It's also a good practice to allow re-use of wasted printed paper next to the copier or in a central location.

### 8.8.4 Eliminating Plastic Water Bottles

Plastic bottles are still used in many office locations to provide water to guests and visitors coming for meetings and other purposes. This can be easily avoided by providing central water dispensers or hydration stations in pantries and other convenient locations (see Fig. 8.12). Some hotels are now practicing this and have managed to eliminate huge amount of plastic waste due to use of plastic bottles for drinking water. Same goes with plastic cups and Styrofoam cups, which are both environmental hazards when they end up in landfills. Providing people with reusable cups in offices can go a long way in reducing plastic waste.

### 8.8.5 Food Waste

Food waste is also a big source of waste in buildings, especially ones where a food court or restaurants are part of the building e.g. shopping malls. Coffee grounds, trimmings from kitchen preparations, partially eaten meals, and leftovers from catering can all contribute to significant amount of organic food waste. Most of the

**Fig. 8.12** Photo of a hydration station that encourages reusable bottles for drinking water

time, the food waste end up in landfills due to a lower recovery of this type of waste in buildings.

As in any waste management program, the best option is to reduce the food wastage by encouraging people to think about the food portions and preparations that are catered to their needs. Food waste can also be avoided by ensuring that the left-over food in a catered buffet can be consumed by others or donated to charity while it's still in good conditions. If food has to be eventually wasted, there are useful ways of handling food waste instead of sending it to landfill or waste incineration. Composting and anaerobic digestion are examples of such useful food waste treatment options.

Composting is method in which food waste can be converted into useful fertilizers by breaking the organic matter down. There are a number of composting types: on-site composting, vermicomposting, aerated composting, and in-vessel composting. Vermicomposting is a solution that is becoming commercially viable. This option uses red worms that are placed in bins with organic matter in order to break it down into a high-value compost called castings. Building could help the composting chain by providing special compost recycle bins for food waste in the building (see Fig. 8.13). Food recycling bins however need to be regularly cleaned in order to ensure that the food waste does not decompose, which may lead to foul smells in the building.

Food waste can also be treated with anaerobic digesters, which consist of a biological process where microorganisms break down biodegradable material in the absence of oxygen. One of the end products of anaerobic digestion is biogas, which is combusted to generate electricity and heat, or can be processed into renewable

**Fig. 8.13** Photo of a recycle bin in an office that as provision to dispose food waste for composting

natural gas and transportation fuels. The main benefit of anaerobic digesters is that they can be self-contained and the food is not exposed to air, which avoids any foul smells during decomposing the food.

## References

AIRAH (2009) Australian Institute if Refrigeration, Air Conditioning and Heating. Water conservation in cooling towers. (online) 140 Johnston St, Fitzroy Victoria 3065. http://www.airah.org.au/imis15_prod/Content_Files/BestPracticeGuides/BPG_Cooling_Towers.pdf. Accessed 30 Dec 2015

Allen L, Smith J, Palaniappan M (2010) The potential of greywater systems to aid sustainable water management, 1st edn. (ebook) 654 13th Street, Preservation Park Oakland, California 94612: Pacific Institute, p 5. https://www.dlsweb.rmit.edu.au/lsu/content/1_studyskills/study_tuts/harvard_ll/direct.html. Accessed 25 Apr 2015

Building and Construction Authority (BCA) of Singapore. Updated April 2016. https://www.bca.gov.sg/BuildableDesign/ppvc.html. Accessed 29 Apr 2016

Caroma USA (2016) http://www.caromausa.com/products/index/cu_products/39.php. Accessed 20 Mar 2016

DrainWorks (2014) 8 ways to conserve water at home, available online at: 8 Ways to Conserve Water at Home. Accessed 18 Nov 2015

EPA (2016) United States Environment Protection agency WaterSense Program, updated Feb 2016. https://www3.epa.gov/watersense/about_us/what_is_ws.html. Accessed 31 Mar 2016

EPD Guidance Document (2007) Georgia Environmental Protection Division Watershed Protection Branch, 1st edn. (ebook) http://www1.gadnr.org/cws/Documents/Leak_Detection_and_Repair.pdf. Accessed 26 Apr 2015

Maimon A, Tal A, Friedler E, Gross A (2010) Safe on-site reuse of greywater for irrigation—a critical review of current guidelines. Environ Sci Technol 44:3213–3220

Rainwater Harvesting Supply Company (2016) System design. http://www.rainharvestmo.com/systemdesign/. Accessed 20 Mar 2016

WaterNSW, Composting Toilets, Dry and Wet. http://www.waternsw.com.au/water-quality/catchment/living/wastewater/systems/composting-toilets. Accessed 21 Feb 2016

WRAP (Water and Resources Action Program) (2016) Designing out waste: a design team guide for buildings. www.wrap.org.uk/construction. Accessed 20 Mar 2016

ZME Science (2013) Waterless urinals: saving water and money from going down the drain. http://www.zmescience.com/space/spacexs-grasshopper-sets-new-record-half-mile-hop-video/. Accessed 25 July 2015

# Chapter 9
# Engaging Occupants in Green and Smart Buildings

The human factor in building technologies cannot be ignored. After all, buildings are made for human occupancy in various forms such as living, working, studying, entertaining, shopping etc. Janda (2011) argues that building users play a critical but poorly understood and often overlooked role in the built environment. Architects, designers and buildings owners should seek ways of integrating user involvement in building performance. One interesting study done by Masoso and Grobler (2010) suggests that more energy is used during non-working hours (56 %) than during working hours (44 %) in office buildings. This arises largely from occupants' behavior of leaving lights and equipment on at the end of the day, and partly due to poor zoning and controls.

Another study by Monfared and Sharples (2011) suggest that occupants need to be associated firmly with the green building provisions as their disengagement with the building's green identity can have a negative impact on the building's sustainable performance. Hence green and smart buildings should view occupants as active participants, both in facilitating comfort and achieving optimal energy efficiency and sustainability.

## 9.1 Increasing Occupant Awareness

The efficacy of behavioral change measures is promising as building studies regularly show that 5–15 % energy savings can be achieved through the provision of sound advice and live feedback on consumption and more if community-scale measures are adopted (Gill 2010). Users need to be provided with tools and information they require to efficiently operate their building. For example, benchmarking comparison to similar buildings means that the excessive users can be motivated to reduce energy and water usage. There are also several studies that

 
N.Y. Jadhav, *Green and Smart Buildings*, Green Energy and Technology,
DOI 10.1007/978-981-10-1002-6_9

show the ‘rebound effect’ where after implementing energy efficiency measures, users start consuming more energy with a mindset that they are now entitled for more.

Various tools such as instructional signage, posters and stickers in strategic locations can provide occupants with tips and information for energy and water savings in green buildings. Some buildings use dashboards and notice boards to convey the message about green buildings and constructive behaviors towards environmental sustainability (see Fig. 9.1). Some of the most recent developments include smart meters, kiosks (e.g. GreenTouchScreen®) and real-time web-based

**Fig. 9.1** Large posters used at a lift lobby in a building in Singapore to make users aware of the green features in the building

feedback (e.g. Building Dashboards) introduced at campuses across North America in an effort to make building performance factors more amenable to understanding and control. Oberlin College, for example, developed an automated data monitoring system to provide dormitory residents with real-time feedback on energy and water use. Students receiving web-based feedback achieved a 55 % reduction in electricity consumption compared to 31 % reduction achieved from meter reading alone (Peterson et al. 2007).

## 9.2 Using the Utility Bill to Engage Occupants

The utility bill is one important and serious touch-point for engaging the occupants as they receive a monthly bill and it is an important document that is often preserved by the occupant or the bill payer. The utility bill could also be used to inform the user of their energy or water consumption relative to their neighbors and other good benchmarks. Figure 9.2 shows a screenshot of the online utility bill portal developed by OPOWER, which an organization specializing in occupant engagement through utility bills and other mechanisms.

One extension of these performance recording systems is the ability to provoke competition among occupants. Competition is always a motivating factor when

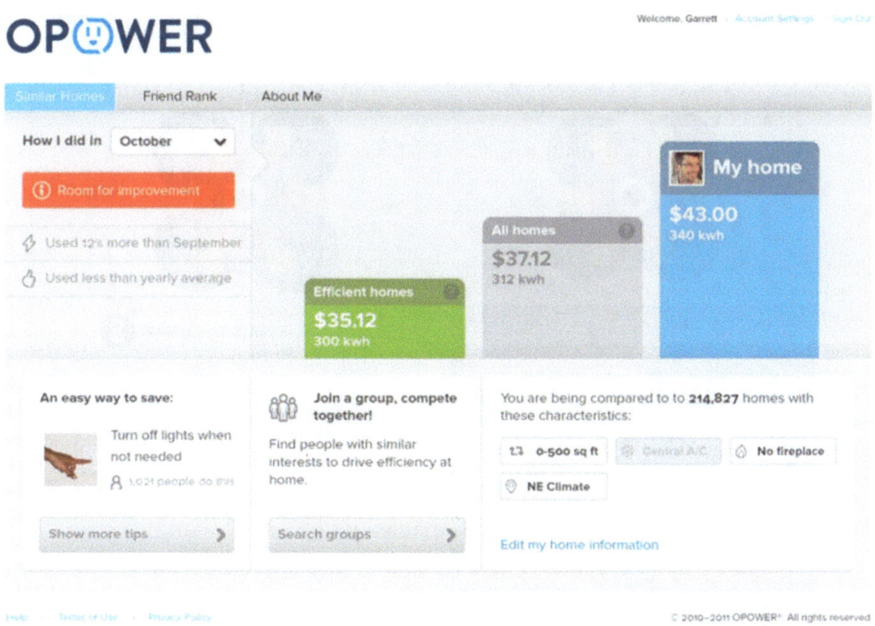


Fig. 9.2 Online utility bill portal that facilitates the comparison of electricity bill with neighbors and other benchmarks (Opower.com)

trying to engage participants. Competition isn't limited to a single office or individual users; it can be among floors of a high-rise, between buildings in a facility, or even between unrelated facilities. With the social networking capabilities, competition can become widespread.

## 9.3 Interactive Applications and Gamification to Engage Occupants

As smartphone is the most common device that anyone has these days, some companies use this medium to reach out to occupants. One such example is the Powerzee smartphone app developed by Engie in collaboration with the Nanyang Technological University (NTU) Singapore. It gives university students and staff the opportunity to make better use of campus resources through an interactive and engaging game app on their smartphone. The app allows users to communicate their comfort conditions (e.g. feeling too hot or too cold in their occupied space) to the facility managers. The data from various users can be then aggregated and some data analytics performed to enable the facility managers to take actions to improve the occupant comfort based on their feedback. Figure 9.3 below shows some screenshots from the Powerzee app.

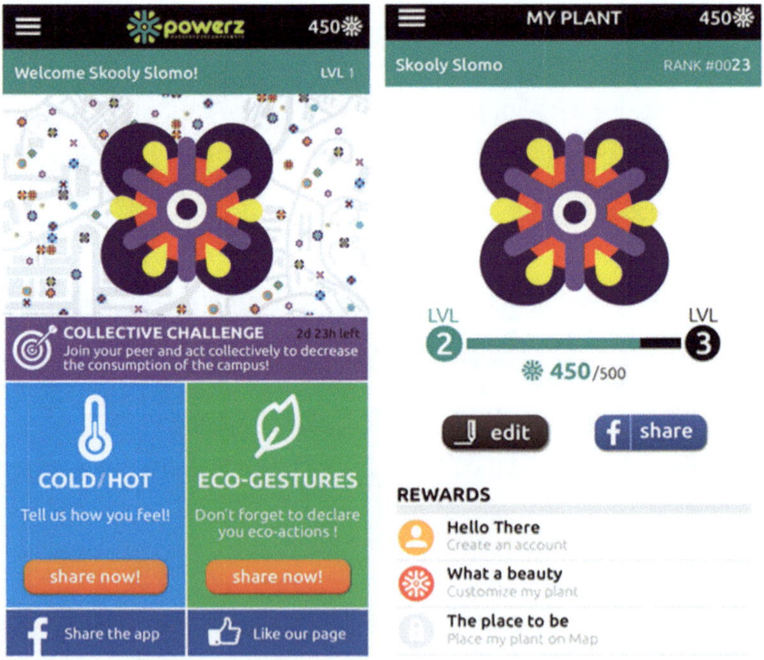


**Fig. 9.3** Screenshots from the Powerzee app (version-1) for energy efficiency through interactive user engagement

The app is presented as a game in which users are asked to indicate their comfort conditions (i.e. feeling too hot/cold) and what they are doing in practical terms to cut their energy consumption (turning off lights, doing without air conditioning, using stairs in preference to elevators, etc.) to win points. The app also suggests eco-friendly actions ('Eco-Gestures') that can optimize day-to-day energy consumption. The ultimate aim is to cut campus energy consumption, water consumption and waste generation. Weekly and monthly analyses that evaluate the impact of these actions on campus energy consumption keep university students and staff informed about the actual impact of their efforts. For more information about this app, please visit: http://www.engie.com/en/news/launch-powerz-eco-responsibility-app/.

There are also similar smartphone apps now available that are informational tools for users to save energy and water in buildings. A list of the apps and their websites is given below for easy reference.

- **Green Outlet**: allows users to determine which home appliances consume the most energy and provide estimated energy bills and possible savings. Available on Apple's iTunes app store.
- **Energy Cost Calculator**: It calculates energy cost of an electronic item based on several equations. The app also shows users their energy usage per day, week, month and year as well as the carbon emission per year. Available on Apple's iTunes app store.
- **Smappee**: In order to use the app, Smappee provides a clip on sensor that users can quickly attach to their fuse box. From there, they can view and monitor their energy usage via the free app. Available online at: http://www.smappee.com/ca_en/energy-monitor.
- **Power Cost Monitor—Energy Cloud**: This app was developed to help homeowners wirelessly monitor their energy consumption and comes with sensors to be installed that provide users with real-time data of their energy usage. Available at: http://www.bluelineinnovations.com/.
- **Wattsly**: advertises itself as a personal energy butler. This app helps people monitor and track their energy consumption by challenging them to lower their energy costs. It's smart graph technology shows usage in high, mid and low peak hours, in an easy to understand format. Available at: http://www.wattsly.com/.
- **Wiser EMS**: Energy management system tool from Schneider Electric. A comprehensive demand management solution for utilities and consumers, allows homeowners to reduce or shift energy use during peak times and helps electricity providers improve grid efficiency and network reliability. Available at: https://www.wiserenergy.com/.

**Table 9.1** Typical examples of provisions made in the green lease agreement

| Green lease provisions | Responsible party |
|---|---|
| Annual targets for energy, water and waste consumption and reduction to be agreed upon and planned for with regular monitoring | Tenant and Landlord |
| Installation of energy and water meter for the tenant space with regular sharing of readings if continuous monitoring is not provided | Landlord |
| Installation of waste quantification measures and recycling bins for waste collection as accessible locations | Landlord |
| Installation of sub-meters to quantify and monitor energy usage within the premise | Tenant |
| Use of energy efficient light fixtures within the tenanted space | Tenant |
| Lighting control to include occupancy sensors and photo sensors (perimeter lighting). Enclosed spaces such as meeting/conference rooms and closed offices to have dedicated light switches and auto switch off when not in use | Tenant (for own space) and landlord (for common areas) |
| Use of water efficient fixtures with certain efficiency rating wherever applicable | Tenant |
| Air conditioning controls to maintain thermal comfort and good indoor air quality shall be provided. Dedicated thermostat controls for enclosed areas | Tenant |
| All control systems and devices (such as sensors and actuators) to be regularly checked and maintained | Tenant |
| Office and other equipment to be bought with certain energy standards rating and the energy efficient or power saving settings be used | Tenant |
| Video conferencing and teleconferencing facilities be provided to avoid excessive travel for business meetings | Tenant and/or Landlord |
| A procurement guideline be developed and followed with clear preference for procuring green materials and equipment | Tenant and Landlord |
| Interior paints, varnishes, sealants and adhesives to be low VOC using natural and water based products where possible | Tenant |
| Internal partitions and fixtures to be modular and re-usable | Tenant |
| Recycle or reuse as much as possible any waste created in the demolition within the premises for refurbishments or alterations | Tenant |
| Encourage staff accountability, awareness and participation in green and smart building provisions | Tenant and Landlord |
| Conduct regular audit for indoor air quality, waste and other energy or water efficiency improvements | Tenant and Landlord |
| The building shall be certified using the known rating systems and comply to all regulatory standards | Landlord |

(continued)

**Table 9.1** (continued)

| Green lease provisions | Responsible party |
|---|---|
| Adequate provision of secured and sheltered bicycle racks and shower facilities for cyclists | Landlord |
| Provision of shuttle bus services to nearby transport node such as train, metro or bus stations if not within walking distance from the building | Landlord |
| Implement a fault reporting and monitoring system to allow staff and tenants to report faults and water leaks and also conduct regular inspections for water leaks and other base building performance issues | Landlord |
| Provide a facility management team with well trained staff with certain minimum number of staff trained for energy/water management and green building standards | Landlord |
| Provide an overview of tenant utility consumption and benchmarks with other tenants. Suggest initiatives and tips to improve operational performance | Landlord |

## 9.4 Green Lease

For commercial places that are rented out, the business case of going green is rather diffused between the different building stakeholders. The landlord or building owner has typically very little say in the fit-outs chosen by the tenant. Also if the building owner decides to do the fit-outs for energy efficiency, the tenant enjoys the benefits at no extra costs. The tenant on the other hand, pays a flat rate based on the area used and may not have a longer term thinking in terms of the space use. Thus the incentive and responsibility for green building features is not clearly assigned in such cases.

Green lease is an agreement between landlord and tenant which sets out environmental objectives on how the building is to be improved, managed and/or occupied in a sustainable manner. The green lease also often splits the rental costs and the utility bill costs by installing a sub-meter for the tenant. This encourages the tenant to use less energy and the building owner is able to pass some of the costs of energy efficiency upgrades to the tenant.

Green lease usually creates a fair and win-win situation for both the parties. For example, by installing energy efficient lighting that generate less heat, a landlord can benefit from a reduced overall air-conditioning energy consumption while a tenant can benefit from the reduced energy bill for lighting usage. The Green Lease improves transparency and accountability through providing a legal agreement in which the parties identify and address problems promptly and efficiently together. Some typical provisions of the green lease agreement that shares the green building operational responsibilities are mentioned in the Table 9.1.

## References

Gill Z (2010) User behaviour key to buildings' emissions. H&V News, August 2010. Available online at: http://www.hvnplus.co.uk/intelligence/user-behaviour-key-to-buildings-emissions/8604489.article. Accessed 30 March 2016

Janda KB (2011) Buildings don't use energy: people do. Architectural Sci Rev 54(1)

Masoso OT, Grobler LJ (2010) The dark side of occupants' behaviour on energy use. Energy Build 42(2):173–177

Monfared IG, Sharples S (2011) Occupants' perceptions and expectations of a green office building: a longitudinal case study. Architectural Sci Rev, 54(4)

Peterson JE, Shunturov V, Janda K, Platt G, Weinberger K (2007) Dormitory residents reduce electricity consumption when exposed to real time visual feedback and incentives. Int J Sustain High Educ 8(1):16–33

# Chapter 10
# Green Building Performance Assessment and Rating

All the technologies and techniques discussed in earlier chapters provide the building stakeholders with enough arsenal for designing and operating a green and smart building. However, one question still remains, and that is about how the performance of green buildings can be measured and assessed to ensure that all that is required is done and how does the building compare with other buildings. Such performance assessment will also provide insights into the opportunities to improve building performance by learning and borrowing ideas from the best practices in green and smart buildings.

However, measuring and benchmarking building performance may not be so straightforward as each building is designed for different functionality and user profiles. The location of the building, its geography and climatic conditions also have an impact on its performance. Also, building performance has to be assessed holistically and not on any single parameter such as energy use intensity. Hence, there has to be a consistent and holistic way to measure and assess building performance. This is what the international rating systems and standards have tried to achieve over the past few years. In this section we discuss the building performance metrics and also the green building rating systems.

## 10.1 Performance Metrics for Green and Smart Buildings

One of the core challenges to achieving widespread adoption of high-performance buildings is to determine which performance metrics are of greatest value to the building community and then to determine the most reliable ways to measure and report them (Deru and Torcellini 2005). Also different stakeholders such as operators, facility managers, designers, building owners and policy makers tend to look at different performance indicators in buildings. In this sections, we look at overall building systems performance measures without going into too much details of individual component performance.



N.Y. Jadhav, *Green and Smart Buildings*, Green Energy and Technology,
DOI 10.1007/978-981-10-1002-6_10

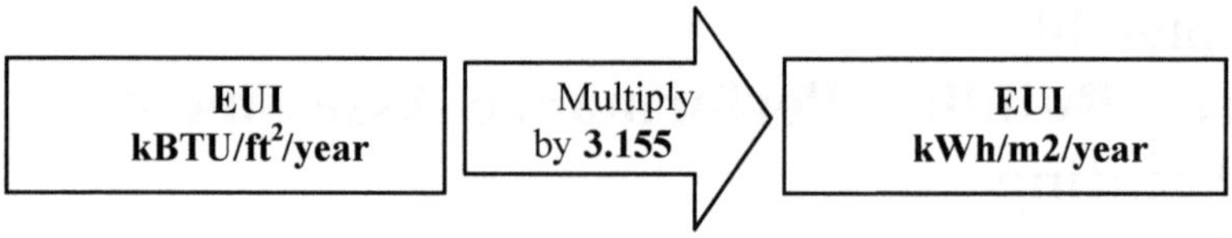


**Fig. 10.1** Conversion between two frequently used units for EUI

### *10.1.1 Energy Use Intensity (EUI)*

The Energy Use Intensity (EUI) or sometimes termed as Energy Utilization Index (EUI) is the one of the most commonly used performance metric in the building industry. The EUI expresses a building's energy use as a function of its size or other characteristics. The most commonly used EUI, expresses the total energy consumed by the building in one year over its total gross floor area (in square meters or square foot). The total energy consumed is usually expressed in kilowatt-hours (kWh) or kilo British thermal units (kBtu) and most of the times can be directly read from the utility bill or the building energy meter (also known as the 'site energy' or the 'final energy'). In some cases, for fair comparison, the term 'source energy' or 'primary energy[1]' is used with buildings that have their own energy generation or heating/cooling plant, versus buildings that import electricity from the grid and heating/cooling from district heating/cooling plants. The units of EUI are typically kBTU/ft$^2$/year or kWh/m$^2$/year and Fig. 10.1 shows a quick conversion between the two units.

Generally, a low EUI signifies good energy performance and most green building designs or retrofits are measured based on their improvements in EUI relative to the base case. Several green building rating systems also use the EUI as the key energy performance indicator to compare building designs with energy efficiency measures and rate them against the base case. However, certain property types will always use more energy than others. For example, an elementary school uses relatively little energy compared to a hospital (EPA 2016). Figure 10.2 is a graph compile by EPA in the United States with data from more than 100,000 buildings reported in their Portfolio Manager (PM) tool.

It is hence important to compare the EUI of buildings with the same building type, based on its basic function. When a building has multiple functions (e.g., office, retail, laboratory, parking), the EUI may be itemized for each functional area for comparison to other buildings of the same types. Alternatively, the building may be analyzed as a whole and reported as a mixed-use building.

Although EUI is a good performance metric for energy efficient and green buildings, its universal applications has several challenges and complexities. For example, sometimes it may not be fair to compare buildings that are densely packed

---

[1]Primary energy is the raw fuel that is burned to create heat and electricity, such as natural gas or fuel oil used in onsite generation.

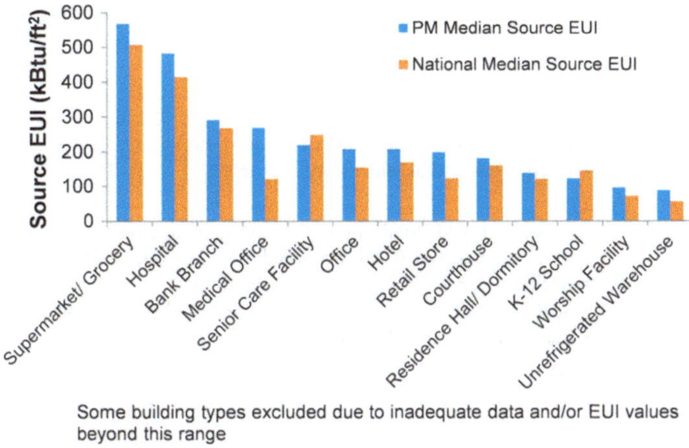


**Fig. 10.2** Example of EUI for different building types. *Source* EPA (2016)

with equipment and people compared to those that are sparsely populated. Hence the total floor area served may not be a good output indicator for the energy use intensity and comparison might be done on basis of occupant population or the economic value created. Also, climate can have a significant effect on EUI, due to the variations in heating and cooling requirements throughout the year. For example, a building in the tropics with cooling requirements cannot be compared at par with a building in a very temperate and mild climate throughout the year. Hence certain normalization parameters have to be used. Table 10.1 shows the different variations possible for measuring energy performance in buildings.

## *10.1.2 Passive Design Performance Metrics*

As discussed in Chap. 4, the main purpose of passive design is to maximize the energy and costs savings by taking advantage of the outside environmental conditions to provide for natural ventilation, natural heating/cooling and day lighting for example. The following metrics can be used to measure the performance of a good passive design that leverages upon layout, massing, orientation and building envelope choices.

### 10.1.2.1 Overall Thermal Transfer Value (OTTV)

In Chap. 4, we touched upon thermal transmittance (U-factor or U-value) of the envelope elements such as the glazing and fenestration in general, opaque surfaces and also the solar heat gain coefficient. The Overall Thermal Transfer Value

**Table 10.1** Variables in assessing and comparing energy performance of buildings (adapted from IEA 2015)

| Type | Parameter |
|---|---|
| Inputs | • Final energy (total, electricity, gas, etc.)<br>• Primary energy (total, electricity, gas, etc.)<br>• Energy cost (total, electricity, gas, etc.) |
| Outputs | *Per*:<br>• Persons served (total population, occupants, employees, etc.)<br>• Floor area served (total, occupied, heated, cooled, enclosed)<br>• Building served (total, grid-connected, etc.)<br>• Service-level provided (amount of heating, cooling, lighting, etc.)<br>• Economic value (GDP, property value, etc.) |
| Scope | *For*:<br>• Sector [all buildings, residential sub-sector, services sub-sector (commercial and public)]<br>• Building types (single-family, multi-family, office, healthcare, hospitality, etc.)<br>• End uses (heating, cooling, water-heating, lighting, appliances, cooking, etc.)<br>• Region (country, state, city, etc.) |
| Normalization factors | *Normalized by*:<br>• Climate (ground temperature, heating degree days, cooling degree days)<br>• Economic indicators (purchasing power parity, currency, etc.)<br>• Time (% change from baseline data, lifecycle) |

(OTTV) is an aggregate calculated value to assess the heat gain from external walls and roof. It will usually consider various components such as:

- heat conduction through opaque walls and roof
- heat conduction through glass windows or skylight
- solar radiation through glass windows or skylight.

The concept of OTTV is based on the assumption that the envelope of a building is completely enclosed and is applicable for mechanically cooled buildings. In some cases, the OTTV is differentiated for the external walls and windows by the term Envelope Thermal Transfer Value (ETTV), and the Roof Thermal Transfer Value (RTTV).

The OTTV is expressed in the units $W/m^2$ and is a reasonable measure of the average heat gain through the roof and envelope of a building at a given location. Some building standards in hot and humid countries especially (such as Singapore and Hong Kong), specify the maximum allowable value for OTTV in design and such codes also specify the calculation methods of OTTV. The lower the OTTV, it's better for reducing the energy used for air conditioning in the building. There are ways of reducing the OTTV of buildings through appropriate combinations of building materials, glass type and architectural modifications.

#### 10.1.2.2 Percentage (%) Naturally Ventilated Spaces

The passive design elements such as layout, massing and orientation can be suitably altered to maximize the natural ventilation in buildings. This avoid the need for mechanical cooling and ventilation, which consume a lot of energy. One measure of a good design in this respect is the percentage (%) of spaces in the building that naturally ventilated. The designer should maximize the percentage naturally ventilated space without compromising comfort and safety.

#### 10.1.2.3 Air Infiltration/Leakage Rate in ACH50 or CFM50

As discussed in Chap. 4 (Sect. 4.4.3), building designs in some climatic conditions (where temperature difference between inside and outside is large) are very sensitive to the air infiltration rate as it can affect the effectiveness of heating and cooling energy use.

The air tightness in buildings is measured by the air changes per hour (ACH), which estimates how many times in one hour the entire volume of air inside the building leaks to the outside. To determine the number of air changes per hour, many experts use the blower door test to create a negative pressure of 50 Pascals. Fifty Pascals is approximately equivalent to a 20 mile per hour wind blowing against all surfaces of the building. The value is then summarized in a metric known popularly as $ACH_{50}$. The other metric that is also often is the cubic feet per minute (CFM) of air moving in the test at 50 pascals of negative pressure (CFM50).

Both CFM50 and ACH50 are used in many building codes and standards to specify the air leakage or infiltration rate and should be as low as possible in a good design. Some standards such as the Passivhaus standard in Europe specify the maximum value of ACH50 in building design. The other measure used in order to allow the air tightness requirement to scale appropriately based on building size is CFM50/$ft^2$ of gross envelope area (Wright and Klingenberg 2015).

#### 10.1.2.4 Daylight Autonomy (DA) and Annual Sun Exposure (ASE)

A good passive design should factor in daylight to minimize the use of artificial lighting during the day and hence save electrical energy use in the building (see Sect. 4.8 for related technologies). Daylight autonomy (DA) is the percentage of the time-in-use that a certain pre-defined lux threshold is reached through the use of just daylight. DA is usually given as an annual value but seasonal, monthly and daily presentations can be made. The calculation procedure of DA is to determine the hourly illuminance level in a point in the room for the entire year. Credit is given if the illuminance level of an hour is above a certain lux threshold and belongs to a time-in-use period of the year. The main advantage of DA is that it can be used to assess the quantitative daylight performance of a design, in contrast to concept of the daylight factor which only deals with the worst case scenario.

One problem with DA is that it is solely a quantitative measure. Furthermore, DA is by definition incremental i.e. it does not give partial credit for daylight levels below the pre-defined lux threshold. This may cause an overestimation of electric lighting energy use for lighting system with daylight control (dimming). hence in some cases, the metric continuous daylight autonomy (cDA) is used. Continuous daylight autonomy (cDA) is similar to DA, but awards partial credit for daylight levels below a pre-defined threshold in a linear fashion.

Another term, viz. Spatial Daylight Autonomy (sDA) describes how much of a space in the building receives sufficient daylight. Specifically, it describes the percentage of floor area that receives at least a pre-defined lux threshold (e.g. 300 lux) of daylight for at least 50 % of the annual occupied hours.

The main problem with sDA is that it contains no upper limit on illuminance levels and hence spaces with too much direct sunlight could appear to do quite well. Hence another counter metric that is Annual Sun Exposure (ASE) is used to check overuse of daylighting provisions. ***Annual Sun Exposure*** (***ASE***) describes how much of space receives too much direct sunlight, which can cause visual discomfort (glare) or increase cooling loads. Specifically, ASE measures the percentage of floor area that receives at least 1000 lux for at least 250 occupied hours per year.

## *10.1.3 Active Design Performance Metrics*

The performance of active design technologies boils down to the efficiency of the equipment used in most cases. In Chap. 5, we have discussed various active design technologies and also touched upon the equipment efficiency metric for heating and cooling equipment (Sect. 5.1), lighting technologies (Sect. 5.2) and other building equipment (Sect. 5.4). Here we quickly recap some of the performance measures discuss in Chap. 5 and discuss some other measures that are typically used in green building standards and rating systems.

### 10.1.3.1 Heating, Cooling and Ventilation Performance Metrics

The most widely used performance metrics for air-conditioning systems are as follows:

(1) Coefficient of Performance (COP): The COP is a measure of the amount of power input to a system compared to the amount of power output by that system. The COP is dimensionless because the input power and output power are measured in Watt. The COP is also an instantaneous measurement in that the units are power which can be measured at one point in time. The higher the COP, the higher the efficiency of the system. In many countries, manufacturers use the refrigeration tons to rate the capacity of the cooling systems and use

the unit kW/ton to express their energy efficiency. The conversion from kW/ton and COP can be done as follows: kW/ton = 3.516/COP.

(2) Energy Efficiency Ratio (EER): The EER is the ratio of output cooling energy (in BTU) to electrical input energy (in Watt-hour). Factoring the right unit conversions, EER = COP $\times$ 3.413.

(3) Seasonal Energy Efficiency Ratio (SEER): The SEER is the same calculation as EER, but the SEER is a representative measurement of how the system behaves over a season where the outdoor temperature and/or humidity varies. Typically, SEER is calculated for the different seasons (4 main seasons in most cases) and then the weighted average is considered to be the overall SEER.

(4) Heating Seasonal Performance Factor (HSPF): It measures the efficiency of the system in heating mode. Therefore, it applies only to heat pumps or reversible air conditioning units and not to units that only cool a space. Like SEER (for cooling mode), this is a measurement of the efficiency of a system and the units are the same as EER or SEER. As with COP, EER, and SEER, the higher the number of HSPF the greater the efficiency. HSPF can also be converted to COP as follows: HSPF = COP $\times$ 3.413

(5) Integrated Part Load Value (IPLV): The Integrated Part Load Value was developed to evaluate the equipment's energy consumption in the cooling mode while it is operating at less than full capacity. The IPLV can essentially be described as a weighted average of the energy efficiency ratio (EER) calculated at each stage of loading at which the unit can operate. The significant difference between them is that the EER is calculated at one very specific point while IPLV is calculated over the unit's entire range of operation. The IPLV can be of significant importance since a unit typically operates at full capacity only a small fraction of the time it is in use.

### *10.1.4 Lighting Power Density*

We looked at several basic lighting performance metrics in Sect. 5.2. However, the main unit that is used to assess overall lighting design performance in green buildings in the Lighting Power Density (LPD). Lighting Power Density (LPD) is defined as watts of lighting per square foot or square meter of room floor area (W/ft$^2$ or W/m$^2$). The lower the LPD, the better as it will translate to significant energy savings depending on the light usage patterns. Using high efficiency lighting (higher lumens per watt) and better daylighting design can improve the lighting power density in green buildings. The lighting power density also affects the air-conditioning performance as the heat emitted from the lights need to be factored in for cooling provisions.

### 10.1.5 *Renewable Energy Performance Metrics*

#### 10.1.5.1 Percentage (%) Energy by Renewables

One of the simplest metric used for green buildings that integrate renewable energy is the fraction expressed in percentage (%) of the total energy used by the building in a year that is be supplied by renewable energy. In case of Zero Energy Buildings, this will 100 %. The measurement for renewable energy production should be normalized to the same units as the total building consumption, for example in kWh/year or Btu/year. In some cases, it's also good to know the renewable energy capacity as % of the peak power requirement. In case of solar PV, the peak generation often also happens during the peak power consumption timeframe during the day and this can provide useful information for the smart grid applications as discussed in Sect. 7.5 earlier.

#### 10.1.5.2 Performance Ratio (PR)

As renewable energy generation sources such as solar and wind tend to be intermittent, not all of the produced power can be fully used in the building at all times. In case of Solar PV, the Performance Ratio (PR) measures how effectively the plant converts sunlight collected by the PV panels into AC energy delivered to the building relative to what would be expected from the panel nameplate rating. This metric quantifies the overall effect of losses due to: inverter inefficiency, wiring, cell mismatch, elevated PV module temperature, reflection from the module front surface, soiling, system down-time, shading, and component failures.

### 10.1.6 *Water-Use Performance Metrics*

Generally, the water use metric for performance assessment and benchmarking is based on either the total gross floor area (GFA, in $m^2$ of $ft^2$) or the number of occupants. Thus the water used is expressed the volume of total water consumed (cubic meters or gallons or litres) per GFA or per person/occupant. The typical interchangeable units that the total water use performance metric is expressed are:

- Gallons/$ft^2$
- Gallons/person
- Cubic meters per $m^2$
- Cubic meters/person.

Apart from the total water usage, if there is sub-metering available to measure water use in different facilities, its good have data for the following water use in the building:

- Indoor potable water use (Gallons/ft$^2$ or Gallons/occupant)
  - Restroom Potable water use (Gallons/ft$^2$ or Gallons/occupant)
  - Kitchen water use (Gallons/ft$^2$ or Gallons/meal)
  - Shower facility water use (Gallons/ft$^2$ or Gallons/occupant)
  - Laundry water use (Gallons/ft$^2$ or Gallons/occupant)
  - Drinking water use (Gallons/ft$^2$ or Gallons/occupant)
- Indoor non-potable water use (Gallons/ft$^2$ or Gallons/occupant)
- Water used for landscaping (Gallons/ft$^2$)
- Cooling tower make-up water use (Gallons/ft$^2$)
- Hot water system make-up water use (Gallons/ft$^2$)
- Rainwater collected (Gallons/time)
- Recycle greywater (Gallons/time)
- Condensate water recovery (Gallons/time).

### *10.1.7 Waste Collection and Recycling Rate*

The weight (in tons) of the total waste collected per person or per GFA is often the metric used for waste. If possible it's also good to know the different source of waste collected such as solid sanitary waste, food waste, hazardous waste, paper and other office materials, building refurbishment/modification and electronic waste.

Other good indicator for waste management is the recycling rate. As the building itself may not have the recycled waste processing capabilities, this is often measured in terms of the total recyclable waste collected from recycling bins. A better indicator for building is the term "diversion rate", which is used to express how much waste is diverted from the landfill. In most cases, the waste is either recycled or composted or reused instead. A green building should have a low total waste generation and a high waste diversion rate.

### *10.1.8 Occupant Satisfaction Metrics*

Occupant satisfaction can be complicated topic as it deals with human behavior and attitude, both of which are dynamic variables that can be not fully linked to the building provision and the occupant's perception can change over time. However, the best way to measure occupant's satisfaction in buildings is to conduct an occupant survey and measure the percentage (%) of satisfied occupants from the total occupants surveyed. Apart from the overall satisfaction result, the survey should also be used to identify critical issues related to occupant comfort and overall functioning and experience in the building.

The survey should be carefully crafted questionnaire and should be detailed enough to understand the key issues related to occupant satisfaction. The following factors could be considered to measure the occupant satisfaction:

- Thermal comfort (temperature, humidity)
- Visual comfort (light levels, glare)
- Auditory comfort (noise levels)
- Olfactory comfort (foul smells)
- Health hazards (volatile organic compounds, carbon monoxide and carbon dioxide levels)
- Visual privacy, conversational privacy
- Turnover rate (% or total)
- Rate of sickness (days per year)
- Absenteeism (days).

### *10.1.9 Life Cycle Costing Analysis*

In the earlier sections we discussed a number if technology options and some of their advantages and disadvantages. When there are a number of alternatives available to fulfill the same performance requirement, the choice is often made based on cost considerations. The Life Cycle Costing Analysis (LCCA) is a method or process of evaluating the cost or economic performance of the building over its entire lifetime. It is sometimes also referred to as 'total cost of ownership' or as 'whole cost accounting'. LCCA is useful in balancing the initial cost of ownership with the long-term expense or savings for the building technologies. The LCCA should be performed early in the design or retrofitting process while there is still a chance to refine the design to ensure a reduction in life-cycle costs. The following costs should be considered in the LCCA for building technologies alternatives over the useful operational life of the building (WBDG 2010):

- Initial Costs—Purchase, Acquisition, Construction Costs
- Fuel Costs
- Operation, Maintenance, and Repair Costs
- Replacement Costs
- Residual Values—Resale or Salvage Values or Disposal Costs
- Finance Charges—Loan Interest Payments
- Non-Monetary Benefits or Cost.

## 10.2 Green Building Standards, Certification and Rating Systems

There is now a proliferation of standards, rating, and certification programs in the marketplace to help guide, demonstrate, and document efforts to deliver sustainable, high-performance buildings. It is estimated that there are nearly 600 green product certifications in the world with nearly 100 in use in the U.S., and the numbers continue to grow. There are also green building rating programs in use around the world and they vary in their approach with some outlining prerequisites and optional credits, while others take a prescriptive approach, and still others suggest performance-based requirements that can be met in different ways for different products and project types (WBDG 2014).

### *10.2.1 Green Building Standards and Codes*

Building energy codes are an important policy instrument to improve energy efficiency in new and existing buildings. In China, for example, residential building energy codes require new buildings to be 65 % more efficient than they were in the early 1980s. The model national codes in the United States require new buildings to be significantly more efficient than buildings constructed even ten years ago. Several different types of building energy codes exist, and often countries will allow multiple pathways for compliance such as prescriptive approach, simulated performance approach, points system and outcome or performance-based codes (IPEEC 2015).

A standard is a set of guidelines and criteria against which a product, design or service can be judged. Common standards related to building practices are created through consensus processes by organizations such as American National Standards Institute (ANSI), American Society for Testing and Materials (ASTM), or American Society of Heating, Refrigerating, and Air-Conditioning Engineers (ASHRAE). Supporting the governance of standards and certifications is the International Standards Organization (ISO), which defines and develops worldwide standards that frequently become law or form the basis of industry norms.

Many of the green product standards available today are proprietary or regulatory standards that have been developed outside of the formal ANSI and ISO consensus process. These types of standards may be more or less stringent than consensus standards and can include some level of transparency and public comment. However, many of these types of standards are trusted because they are associated with a group that has strong environmental credentials.

One such example of green building standards is the ANSI/ASHRAE/USGBC/IES Standard 189.1, "Standard for the Design of High Performance Green Buildings except Low-Rise Residential Buildings" provides minimum requirements for site, design, construction and operations in mandatory, code-enforceable language. This standard is comprehensive and includes chapters for site, water, energy efficiency, indoor environmental quality, and materials.

***The International Green Construction Code*** (IGCC) is intended to be used as a jurisdictional and municipal building code for new construction and major renovations. The IGCC is a comprehensive code document setting standards for energy conservation, water efficiency, and commissioning, and also includes enforcement procedures and guidelines for existing building renovations (ICC 2016).

***The Passive House Standard*** is a leading standard in energy efficient construction. The Passive House Standard stands for quality, comfort and energy efficiency. Passive Houses require very little energy to achieve a comfortable temperature year round, making conventional heating and air conditioning systems obsolete. While delivering superior levels of comfort, the Passive House Standard also protects the building structure (iPHA 2016). or a building to be considered a Passive House, it must meet the criteria as shown in Table 10.2 (for detailed criteria, please see www.passivehouse.com).

## *10.2.2 Green Building Certification and Rating*

Both standards and product certifications may play a role in determining the level of sustainability or performance of a product or equipment. However, each must be considered as part of a larger process of integrating them into the overall project goals to ensure the entire building is sustainable. Green building rating or certification systems broaden the focus beyond the product to consider the project as a whole. Rating systems are a type of building certification system that rates or rewards relative levels of compliance or performance with specific environmental goals and requirements. Rating systems and certification systems are frequently used interchangeably. A few of these programs are single-attribute, focusing solely on water or energy, while others are multi-attribute addressing emissions, toxicity, and overall environmental performance in addition to water and energy. While the philosophy, approach, and certification method vary across these the systems, a common objective is that projects awarded or certified within these programs are designed to reduce the overall impact of the built environment on human health and the natural environment.

### 10.2.2.1 Energy Star

Energy Star is a single attribute international rating and certification system that focuses on energy and water use in buildings. It is managed by the United States (US) Environmental Protection Agency (EPA) and Department of Energy (DOE) but is now adopted internationally in countries such as Australia, Canada,

**Table 10.2** Key requirements of the passive house standard (Passivehouse.com)

| Criteria | Requirement |
|---|---|
| Space heating demand | Not to exceed 15 kWh annually OR 10 W (peak demand) per square metre of usable living space |
| Space cooling demand | Roughly matches the heat demand with an additional, climate-dependent allowance for dehumidification |
| Primary energy demand | Not to exceed 120 kWh annually for all domestic applications (heating, cooling, hot water and domestic electricity) per square meter of usable living space |
| Airtightness | Maximum of 0.6 ACH50 i.e. air changes per hour at 50 Pascals pressure (as verified with an onsite pressure test in both pressurized and depressurized states) |
| Thermal comfort | Thermal comfort must be met for all living areas year-round with not more than 10 % of the hours in any given year over 25 °C |

Japan, New Zealand, Taiwan, and the European Union. Products are certified using a benchmarking method and the Energy Star service mark (as shown in Fig. 10.3) is placed on energy-efficient products.

The Energy Star program has also developed energy performance rating systems for several commercial and institutional building types and manufacturing facilities. These ratings, on a scale of 1–100, provide a means for benchmarking the energy efficiency of specific buildings and industrial plants against the energy performance of similar facilities. The ratings are used by building and energy managers to evaluate the energy performance of existing buildings and industrial plants. The rating systems are also used by EPA to determine if a building or plant can qualify to earn Energy Star recognition.

Energy Star has also created an online tool called the Energy Star Portfolio Manager that helps to track energy use, water use and greenhouse gas emissions of buildings. The database of buildings across the world is used for benchmarking green building performance. This online tool is available at: www.energystar.gov/benchmark.

#### 10.2.2.2 LEED Certification

The LEED (Leadership in Energy and Environmental Design) certification is one of the most popular green building rating and certification system worldwide. It is managed by the United States Green Building Council (USGBC) and provides for green building rating and certification through independent third-party verification. The following areas are covered in the LEED rating and certification system (WBDG 2014):

1. Sustainable Sites
2. Water Efficiency
3. Energy and Atmosphere
4. Materials and Resources

**Fig. 10.3** The Energy Star logo found on building products signifies that they are energy and/or water efficient

5. Indoor Environmental Quality
6. Locations and Linkages
7. Awareness and Education
8. Innovation in Design
9. Regional Priority: geographically specific environmental priorities.

The LEED rating can now be applied to five different building project types in various building functionality and phases as follows (USGBC 2016):

(a) Building Design + Construction: Applies to buildings that are being newly constructed or going through a major renovation; includes New Construction, Core and Shell, Schools, Retail, Hospitality, Data Centers, Warehouses and Distribution Centers, and Healthcare.
(b) Interior Design + Construction: Applies to projects that are a complete interior fit-out; includes Commercial Interiors, Retail and Hospitality.
(c) Building Operations + Maintenance: Applies to existing buildings that are undergoing improvement work or little to no construction; includes Existing Buildings, Schools, Retail, Hospitality, Data Centers, and Warehouses and Distribution Centers.
(d) Neighborhood Development: Applies to new land development projects or redevelopment projects containing residential uses, nonresidential uses, or a mix. Projects can be at any stage of the development process, from conceptual planning to construction; includes Plan and Built Project.
(e) Homes: Applies to single family homes, low-rise multi-family (one to three stories), or mid-rise multi-family (four to six stories); includes Homes and Multifamily Low-rise and Multifamily Midrise.

LEED is a multi-level (certified, silver, gold and platinum levels) rating system based on credit points in different categories along with mandatory prerequisites such as minimum energy and water-use reduction, recycling collection, and tobacco smoke control that can be specified at each level. Within each category are credits that pertain to specific strategies for sustainability, such as the use of low-emitting products, reduced water consumption, energy efficiency, access to public transportation, recycled content, renewable energy, and daylighting.

#### 10.2.2.3 Living Building Challenge (LBC)

The Living Building Challenge (LBC) is a performance-based system initially launched by the Cascadia Green Building Council. In April 2011, the International Living Future Institute became the umbrella organization for both the Cascadia Green Building Council and the Living Building Challenge.

The Challenge is comprised of seven performance categories called Petals: Place, Water, Energy, Health and Happiness, Materials, Equity and Beauty. Petals are subdivided into a total of twenty Imperatives, each of which focuses on a specific sphere of influence. This compilation of Imperatives can be applied to almost every conceivable building project, of any scale and any location—be it a new building or an existing structure.

The LBC makes stringent demands such as 100 % net zero energy, 100 % net zero water, on-site renewable energy, and 100 % recycling or diversion of construction waste. It examines site, water, energy, materials, health, equity, and beauty. All of its tenets are mandatory making it the most rigorous green building certification system in the market today. For more information, visit: http://living-future.org/lbc.

#### 10.2.2.4 International Certification Programs

Apart from the above popular programs, there several regional or country-led green building certification programs that are used for certifying Green and Smart Building. Some of the well-known regional programs are listed below.

- BREEAM: UK, EU, EFTA member states, EU candidates, as well as the Persian Gulf.
- CASBEE: Japan
- Green Mark Scheme: Singapore
- BEAM: Hong Kong
- Pearl Rating System for Estidama: UAE.

Each of these rating systems have common as well as different elements or criteria of sustainability performance. Often, the emphasis or weightage assigned to a particular criterion and the pre-requisite conditions varies between the different rating systems, considering the local context and priorities. For example, the energy based criteria have an overall weightage of more than 60 % in Singapore's Green Mark scheme, whereas it has less than 30 % weightage in the LEED rating system. The choice of the rating systems tends to have a regional bias as most building owners and facility managers would like to support their local government initiatives. However, other factors such as the scope of issues handled, targets, international outreach and client preferences also influence the choice of the right green building rating system.

## References

Deru M, Torcellini P, Performance metrics research project—Final report (2005) Technical report NREL/TP-550–38700

EPA (Environmental Protection Agency of the USA), ENERGY STAR website, accessible online at: https://www.energystar.gov/buildings/facility-owners-and-managers/existing-buildings/use-portfolio-manager/understand-metrics/what-energy. Accessed 12 Apr 2016

ICC (International Code Council), International Green Construction Code, Available online from: https://www.iccsafe.org/codes-tech-support/international-green-construction-code-igcc/international-green-construction-code/ Accessed 28 Feb 2016

IEA (International Energy agency) (2015) Building energy performance metrics—Building energy performance metrics, OECD/IEA

IPEEC International Partnership for Energy Efficiency Corporation), Delivering Energy Savings in Buildings, IPEEC Building Energy Efficiency Task group Report 2015. Available online at: http://www.gbpn.org/sites/default/files/Delivering%20Energy_full%20report.pdf. Retrieved on 21 Apr 2016

iPHA (International Passive House Association), Passive House certification criteria, Available online from: http://www.passivehouse-international.org. Accessed 29 Feb 2016

USGBC (United States Green Building Council), LEED program, available online from: http://www.usgbc.org/leed. Accessed 21 Mar 2016

WBDG (Whole Building Design Guide), Green Building Standards and Certification Systems, Updated 27 October 2014. Available online from: https://www.wbdg.org/resources/gbs.php. Accessed 10 Apr 2016

WBDG (Whole Building Design Guide), Life-Cycle Cost Analysis (LCCA), Updated 28 June 2010. Available online from: https://www.wbdg.org/resources/lcca.php. Accessed 14 Apr 2016

Wright GS, Klingenberg K (2015) Climate-specific passive building standards, Report prepared for The National Renewable Energy Laboratory on behalf of the U.S. Department of Energy

# Chapter 11
# Conclusion

Green and Smart Buildings technologies discussed and presented in this book are not necessarily all new and cutting edge technologies. Some of them have been available for decades. However, the adoption rates of these technologies are still very low. The list of the technologies discussed in this book, although not exhaustive, can offer a quick reference guide for building owners, designers, architects and facility managers to design and operate green and smart buildings. If the technologies discussed here can be adopted for new buildings and building retrofits, the energy, water and waste footprint of the built environment can be significantly reduced, while at the same time ensuring occupant comfort and safety with the appropriate choice of technologies.

The best time to plan and create green and smart buildings is during the early design stages of the building. It's important that various building stakeholders (building owner, architect, engineers, designers, operators, consultants and users) are involved in the design stage of the building in the integrated design process as discussed in Chap. 3. During this process it is very important to use the right modeling and simulation tools to assess building performance using different ideas and options that are discussed during the process. These tools such as Building Information Modeling (BIM), energy modeling and air-flow modeling using computational fluid dynamics (CFD) can go a long way in documenting the characteristics and performance of the building and provide transferable insights on the choice of technologies and their overall performance impacts.

A lot can be done with the 'passive' design elements of the building such as its orientation, layout and the choice of building envelope or the skin of the building. Passive design can have a major impact on the building performance by affecting its ventilation, lighting and heating and cooling needs. By leveraging the natural environmental factors, the passive design technologies discussed in Chap. 4 can help set up the building for its environmental performance and occupant comfort. It is highly recommended that the passive design of the building be thoroughly optimized first as once fixed, it is often not so easy to change the passive design elements and definitely not at low costs.



N.Y. Jadhav, *Green and Smart Buildings*, Green Energy and Technology,
DOI 10.1007/978-981-10-1002-6_11

After optimizing the building using passive design technologies, the active design provisions of the buildings should be appropriately chosen to ensure its high performance. The active design elements include the heating, cooling and ventilation, artificial lighting, building services such as elevators and the plug and process loads. In Chap. 5 we discussed energy efficient technologies in each of this areas and there is a lot of potential for energy and operational cost savings in the building by choosing the suitable technologies in this section. Active design technologies should also be investigated at the time of the building retrofit and refurbishment. They can offer good solutions for energy efficient retrofits and help the building achieve higher performance after retrofit and refurbishment.

The Building Management System (BMS) or Building Automation System (BAS) helps the building maintain its operational performance. Tools such as continuous monitoring, data analytics, dashboards, sensors and controls can help the facility managers, building owners and occupants to optimize its energy and water efficiency and reduce wastage. Several building provisions and their efficient operation can be automated by the BAS and use of advanced techniques such as automated fault detection and diagnosis can greatly help the facility managers to optimize the building for its highest overall performance at all times. The combination of Internet of Things (IoT) technologies in this area can offer low cost solutions to significantly enhance the 'smartness' of the building and drastically improve the overall occupant experience.

The integration of renewable energy generation options discussed in Chap. 7 as well as the water and waste reduction technologies discussed in Chap. 8, would enable buildings to holistically achieve its next level of sustainability performance. However, without engaging the occupant in the right way, the technologies may fail to deliver their intended purpose. Hence its crucial to focus on occupant engagement using some of the technologies discussed in Chap. 9. In Chap. 10 we looked at performance assessment and rating, which enables the building to benchmark itself against best practices and also motivates and guides the building stakeholders to critically think about the performance of their building and the alternative technologies available.

In a post-occupancy audit and survey conducted by the United States, General Services Administration (GSA), it was found that green buildings that incorporate sustainable practices and are rated as such using the green building rating system LEED, are indeed able to achieve the following compared to national average:

- significant reduction in energy consumption: 25 % less energy use and 58 % savings in energy costs in case of top performers.
- reduced water consumption: 11 % reduction on average.
- decrease carbon dioxide emissions: 36 %.
- higher occupant satisfaction: 27 % on average and 76 % higher in case of top performers.
- lower aggregate operational costs: 19 % lower on average and 43 % in case of top-performers.

- lower aggregate maintenance costs: 12 % lower on average and 47 % in case of top performers.

The survey was conducted on 22 representative green buildings from the GSA national portfolio and involved a comprehensive assessment, measuring environmental performance, financial metrics, and occupant satisfaction. Results were compared to both industry and GSA baselines. It also found that the Integrated Design approach (as discussed in Chap. 3 here) delivers higher performance and help meet the national climate change and sustainability goals (GSA 2011).

There are several examples of green and smart buildings in the world now that have achieved a very high energy and sustainability performance. As of last year (August 2015), there are more than 72,500 LEED certified green building projects located in over 150 countries and territories. A review of data from 195 LEED projects found that green buildings have a 57 % lower source Energy Use Intensity (EUI) than the national average (USGBC 2015).

One good example of a green and smart building in the **Bullitt Centre building** in Seattle, Washington in the United States of America (photo shown in Fig. 11.1). Touted as America's greenest office building by some (BuildingGreen 2016), this six-storey, 52,000 square-foot building completed in April 2013, can meet the Living Building Challenge, which is one of the most stringent challenges for green buildings as discussed in Chap. 10. It met the specific requirements of the challenge such as the net-zero-energy, net-zero-water and sustainable construction materials. Some of the features of this building include the following (WBDG 2016):

- *Integrated design approach* (see Chap. 3): The project made good use of a substantial pre-design phase, where building size and massing, architectural and MEP systems, and renewable energy production potential were proven, prior to the start of schematic design. It also made extensive use of Building Information Modeling (BIM) and energy modeling tools to support design decisions.
- *Passive Design* (see Chap. 4): Daylighting analysis drove not only the massing of the project, but the configuration of the curtain walls, skylights, and shading. Wherever possible, wood (timber) was used resulting in a hybrid structure consisting of concrete coming out of the earth, steel to resist lateral forces and timber for gravity loading conditions. It has a highly superior building envelope performance with use of triple-glazing windows, well insulated walls and effective massing and orientation for reducing solar heat gain and improve daylighting. It also incorporates automated exterior shading and operable windows that open and close automatically in response to conditions outside (a manual over-ride is included for occupant choice).
- *Active design* (see Chap. 5): It incorporates radiant floor heating and cooling coupled with a ground-source heat pump (GSHP) system. There is also a heat recovery system that system includes heat exchangers made of honeycombed rotating drums to extract heat from the warm air before it's vented outside. A very good lighting power density (LPD) of 0.4 W/ft$^2$ (or 4.3 W/m$^2$) is achieved with good daylight planning (2 % daylight factor) and use of LED

**Fig. 11.1** Photo of the Bullitt Centre building (Alex Wilson, BuildGreen 2016)

lighting that has automatic sensor-based control for dimming and on/off features based on occupancy and daylight. The elevators used have a regenerative drives system that makes them much more energy efficient. Plug loads for office equipment, such as computers, monitors, servers, printers, and copiers are limited to a maximum of 0.8 W/ft$^2$ and further significantly reduced by using plug load occupancy sensors.

- *Building Management and Automation System* (see Chap. 6): The Bullitt Center's sophisticated Building Management System (BMS) includes sensors and actuators to monitor and control its heating, cooling, and ventilation systems. It also manipulates external louvers and the operable windows. In addition, the BMS monitors the greywater system and controls the composting system as well. The building employs an online dashboard that shows real-time energy usage and other building functions.

- *Renewable Energy Integration* (see Chap. 7): A 242 kilowatt peak (kWp) rooftop solar photovoltaic (PV) system is installed in the building to deliver 100 % of the building's electricity needs on an annual basis. To get a large enough PV array on the roof to supply electricity for six floors, the PV array cantilevers out over the walls (as can be seen in its photo above). The building also uses the geothermal heating and cooling system that consist of cross-linked polyethylene (PEX) tubes in the floors that carry a mixture of water and glycol throughout the building and 120 m (400 ft) underground.
- *Water and waste reduction* (see Chap. 8): The building has a 56,000 gallon (212 cubic meters) cistern for storage of rainwater that is harvested on the roof and meets 100 % of the water need in the building for drinking, sanitation and landscaping irrigation. The building has composting toilets, which divert the toilet waster directly to composting units located in the building basement. The toilets also use foam flushing system with natural soaps to reduce the water use per flush by a significant amount. The third floor terrace of the building serves as its greywater filtration system (1500 L or 400 gallons per day) that consists of a man-made wetland with horsetail plants that absorb organic material and purify the water.
- *Occupant engagement* (see Chap. 9): There is green lease in place that includes penalties if their energy budgets are exceeded by the tenants. There is an a unique, internal "cap-and-trade" system in which tenants have specific energy budgets and if they use less electricity than their budget, they can trade with other tenants in the building who may need more. Lease incentives for the tenants are used to ensure receptacle energy targets are met and to encourage tenants to employ the most efficient state-of-the-art equipment that meets their professional needs. Large interpretive and educational banners placed in the open reception area of the building show-off the green features of the building to occupants and visitors.
- *Overall performance* (see Chap. 10): The building was designed to achieve an overall energy utilization index (EUI) of 16 kBtu/ft$^2$/year (or 51 kWh/m$^2$/year). The onsite renewable energy generation (from the solar PV panels) of 257,800 kWh/year, meets slightly more than 100 % of the building's annual energy use (BuildingGreen 2016). The following chart in Fig. 11.2 shows the annual energy split by end use for the building.

Another green and smart building case worth mentioning is that of **The Edge building** in Amsterdam, the Netherlands (Photo shown in Fig. 11.3). This building is considered as the greenest and smartest building in the world (Bloomberg 2015). The Edge is a 40,000 m$^2$, 15 floors office building in Amsterdam that was designed

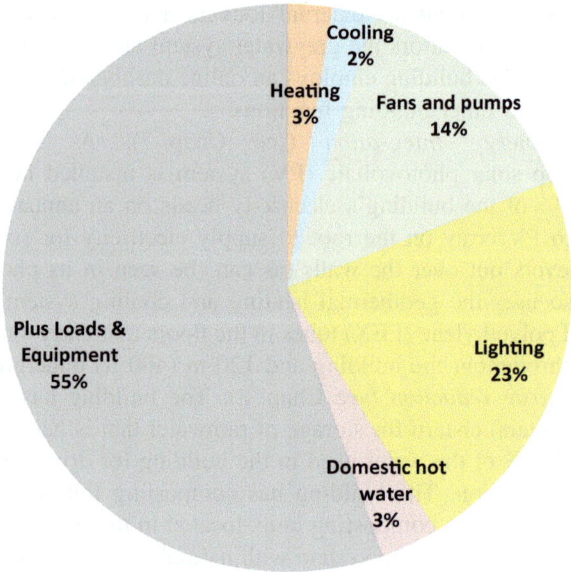


**Fig. 11.2** Annual energy consumption split by end-use for the Bullitt Centre building

**Fig. 11.3** Photo of the edge building

for the global financial firm and main tenant, Deloitte. In the year 2016, it was awarded the world's highest BREEAM rating awarded to an office building ('outstanding' rating with a score of 98.36 %) as it demonstrates that the pursuit of a vibrant and collaborative working environment can be combined successfully with achieving the highest levels of sustainability (BREEAM 2016). The building combines the following green and smart building features:

- *Passive design* (see Chap. 4): The building orientation is based on the path of the sun and harnesses northern daylight while there are solar panels installed on the south facade to harness sunlight for energy generation. Load bearing walls to the south, east and west have smaller openings to provide thermal mass and shading, and solid operable panels for ventilation. Louvers on the south facades are designed according to sun angles and provide additional shading for the office spaces, reducing solar heat gain. Every workspace is planned within 7 m (23 ft) of a window, allowing occupants to enjoy outside views as well as daylight.
- *Active design* (see Chap. 5): The spaces are heated and cooled using radiant cooling ceiling panels that are combined with a heat pump and an aquifer thermal energy storage system that provides all of the energy required for heating and cooling. During summer months, the building pumps warm water more than 400 ft deep in the aquifer beneath the building, where it sits, insulated, until winter, when it's sucked back out for heating. The atrium acts as a buffer between the workspace and the external environment. Excess ventilation air from the offices is used again to air condition the atrium space. The air is then ventilated back out through the top of the atrium where it passes through a heat exchanger to make use of any warmth. The building has an Ethernet-powered LED lighting system, whereby the lighting is powered using the same cables that carry data for the Internet.
- *Building Management and Automation System* (see Chap. 6): Almost all equipment and devices in the building, including the coffee machine and the towel rails in bathroom, are connected to the internet. The building has a newly developed lighting system called Light over Ethernet (LoE). The LED luminaries are not only powered by the Ethernet cables, but also fully internet connected. They have embedded sensors to continuously measure occupancy, movement, lighting levels, humidity and temperature, and then provide that data to the Building Management System (BMS). There are over 28,000 sensors in the building's smart ceiling and lighting system. There is a central dashboard that can process a massive amount of data collected in the building and is accessible to the facility managers and occupants. Heating, cooling, fresh air and lighting are fully IoT (Internet of Things) integrated and BMS controlled per 200 square feet based on occupancy to ensure that with zero occupancy there is next-to-zero energy use. On days when fewer employees are expected, an entire section might even be shut down, cutting the costs of heating, cooling, lighting, and cleaning.

- *Renewable Energy Integration* (see Chap. 7): 65,000 square feet of solar PV panels are located on the facades and roof, and remotely on the roofs of buildings of the University of Amsterdam, thereby making use of neighborhood level energy sourcing. This makes the building net-zero energy building and in fact positive energy as the solar panels produce much more energy than what is consume in the building.
- *Water and waste management* (see Chap. 8): Rain water is collected on the roof, stored in a concrete tank back of the parking garage, and used to flush toilets, and irrigate the green terraces in the atrium and other garden areas surrounding the building. The connected nature of its infrastructure also provides data to be able to reduce wastage in many areas. For example, the data from sensors is also used for predictions of occupancy at lunchtime based on real time historical data and traffic and weather information to avoid food-waste.
- *Occupant engagement* (see Chap. 9): This is an area where the building surpasses all expectations on a green and smart building. Every employee is connected to the building via an app on their smart phone. Using the app, they can find parking spaces, free desks and lockers (as no one has a pre-assigned desk) or other colleagues, report issues to the facilities team, or even navigate within the building. Employees can customize the temperature and light levels anywhere they choose to work in the building via the mobile app. The app checks the occupant's schedule, and the building recognizes their car when they arrive and directs them to a parking spot. The app also remembers how they like their coffee, and tracks their energy use so they're aware of it.

The case study of these two high performance buildings tells us that the technology necessary to create green and smart buildings is now available. There will always be further technological advancement, but with the use of existing technologies it's possible to design and operate buildings that can have a very low or even zero energy, water and waste footprint. The integrated design process used to bring together the several stakeholders in the building should also be used as a platform to generate new ideas and trigger research requirements to enhance building performance even further. Each building is unique based on its location, surrounding climatic conditions and its intended end-use. However, as is evident from the case studies, the creation of green and smart buildings of the future needs people who are aware of the benefits, technologies and processes to make them a reality.

It is a sincere hope that this book triggers interest of the building stakeholders to start championing green and smart buildings and accept nothing less. This book could be used a quick reference to understand, select and prioritize technologies that could be applied to new building designs and retrofits. The technologies described in this book could also be subject of future research and development in order to make them suitable for the local context and also make them cost-effective and readily adoptable by the building industry.

# References

Bloomberg (2015) The smartest building in the world-inside the connected future of architecture, Bloomberg Businessweek, 23 September 2015. Available online at: http://www.bloomberg.com/features/2015-the-edge-the-worlds-greenest-building/. Accessed 16 Mar 2016

BREEAM (2016) Case study of the edge building. Available online at: http://www.breeam.com/index.jsp?id=804. Accessed 30 Apr 2016

BuildingGreen (2016) America's greenest office building. Online article available at: https://www.buildinggreen.com/news-article/americas-greenest-office-building. Accessed 30 Mar 2016

GSA (General Services Administration of the United States of America) (2011) Green building performance—a post occupancy evaluation of 22 GSA Buildings, White Paper in August 2011. Available online from: http://www.gsa.gov/graphics/pbs/Green_Building_Performance.pdf. Retrieved 30 Mar 2016

USGBC (United States Green Building Council) (2015) Green building facts, 23 February 2015. Available online from: http://www.usgbc.org/articles/green-building-facts. Accessed 30 March 2016

WBDG (Whole Building Design Guide) (2016) Available online at: https://www.wbdg.org/references/casestudies.php. Accessed 30 Mar 2016

MIX
Papier aus verantwortungsvollen Quellen
Paper from responsible sources
FSC® C105338

If you have any concerns about our products,
you can contact us on
**ProductSafety@springernature.com**

In case Publisher is established outside the EU,
the EU authorized representative is:
**Springer Nature Customer Service Center GmbH**
**Europaplatz 3, 69115 Heidelberg, Germany**

Printed by Libri Plureos GmbH
in Hamburg, Germany